These safety symbols are used in laboratory and field investigations in tl
ing of each symbol and refer to this page often. *Remember to wash you*

PROTECTIVE EQUIPMENT   Do not begin any lab without the

GOGGLES Proper eye protection must be worn when performing or observing science activities that involve items or conditions as listed below.

APRON Wear an approved apron when using substances that could stain, wet, or destroy cloth.

SOAP Wash hands with soap and water before removing goggles and after all lab activities.

GLOVES Wear gloves when working with biological materials, chemicals, animals, or materials that can stain or irritate hands.

LABORATORY HAZARDS

Symbols	Potential Hazards	Precaution	Response
DISPOSAL	contamination of classroom or environment due to improper disposal of materials such as chemicals and live specimens	• DO NOT dispose of hazardous materials in the sink or trash can. • Dispose of wastes as directed by your teacher.	• If hazardous materials are disposed of improperly, notify your teacher immediately.
EXTREME TEMPERATURE	skin burns due to extremely hot or cold materials such as hot glass, liquids, or metals; liquid nitrogen; dry ice	• Use proper protective equipment, such as hot mitts and/or tongs, when handling objects with extreme temperatures.	• If injury occurs, notify your teacher immediately.
SHARP OBJECTS	punctures or cuts from sharp objects such as razor blades, pins, scalpels, and broken glass	• Handle glassware carefully to avoid breakage. • Walk with sharp objects pointed downward, away from you and others.	• If broken glass or injury occurs, notify your teacher immediately.
ELECTRICAL	electric shock or skin burn due to improper grounding, short circuits, liquid spills, or exposed wires	• Check condition of wires and apparatus for fraying or uninsulated wires, and broken or cracked equipment. • Use only GFCI-protected outlets	• DO NOT attempt to fix electrical problems. Notify your teacher immediately.
CHEMICAL	skin irritation or burns, breathing difficulty, and/or poisoning due to touching, swallowing, or inhalation of chemicals such as acids, bases, bleach, metal compounds, iodine, poinsettias, pollen, ammonia, acetone, nail polish remover, heated chemicals, mothballs, and any other chemicals labeled or known to be dangerous	• Wear proper protective equipment such as goggles, apron, and gloves when using chemicals. • Ensure proper room ventilation or use a fume hood when using materials that produce fumes. • NEVER smell fumes directly. • NEVER taste or eat any material in the laboratory.	• If contact occurs, immediately flush affected area with water and notify your teacher. • If a spill occurs, leave the area immediately and notify your teacher.
FLAMMABLE	unexpected fire due to liquids or gases that ignite easily such as rubbing alcohol	• Avoid open flames, sparks, or heat when flammable liquids are present.	• If a fire occurs, leave the area immediately and notify your teacher.
OPEN FLAME	burns or fire due to open flame from matches, Bunsen burners, or burning materials	• Tie back loose hair and clothing. • Keep flame away from all materials. • Follow teacher instructions when lighting and extinguishing flames. • Use proper protection, such as hot mitts or tongs, when handling hot objects.	• If a fire occurs, leave the area immediately and notify your teacher.
ANIMAL SAFETY	injury to or from laboratory animals	• Wear proper protective equipment such as gloves, apron, and goggles when working with animals. • Wash hands after handling animals.	• If injury occurs, notify your teacher immediately.
BIOLOGICAL	infection or adverse reaction due to contact with organisms such as bacteria, fungi, and biological materials such as blood, animal or plant materials	• Wear proper protective equipment such as gloves, goggles, and apron when working with biological materials. • Avoid skin contact with an organism or any part of the organism. • Wash hands after handling organisms.	• If contact occurs, wash the affected area and notify your teacher immediately.
FUME	breathing difficulties from inhalation of fumes from substances such as ammonia, acetone, nail polish remover, heated chemicals, and mothballs	• Wear goggles, apron, and gloves. • Ensure proper room ventilation or use a fume hood when using substances that produce fumes. • NEVER smell fumes directly.	• If a spill occurs, leave area and notify your teacher immediately.
IRRITANT	irritation of skin, mucous membranes, or respiratory tract due to materials such as acids, bases, bleach, pollen, mothballs, steel wool, and potassium permanganate	• Wear goggles, apron, and gloves. • Wear a dust mask to protect against fine particles.	• If skin contact occurs, immediately flush the affected area with water and notify your teacher.
RADIOACTIVE	excessive exposure from alpha, beta, and gamma particles	• Remove gloves and wash hands with soap and water before removing remainder of protective equipment.	• If cracks or holes are found in the container, notify your teacher immediately.

Your online portal to everything you need

connectED.mcgraw-hill.com

Look for these icons to access exciting digital resources

- Video
- Audio
- Review
- Inquiry
- WebQuest
- Assessment
- Concepts in Motion

McGraw Hill Education

EXPLORING THE UNIVERSE

iSCIENCE

Glencoe

Geode
This is a cross section of geode, a type of rock. The outside of a geode is generally limestone, but the inside contains mineral crystals. Crystals only partially fill this geode, but other geodes are filled completely with crystals.

Copyright © 2012 The McGraw-Hill Companies, Inc. All rights reserved. No part of this publication may be reproduced or distributed in any form or by any means, or stored in a database or retrieval system, without the prior written consent of The McGraw-Hill Companies, Inc., including, but not limited to, network storage or transmission, or broadcast for distance learning.

Send all inquiries to:
McGraw-Hill Education
8787 Orion Place
Columbus, OH 43240-4027

ISBN: 978-0-07-888012-4
MHID: 0-07-888012-2

Printed in the United States of America.

8 9 10 11 LKV 19

Authors and Contributors

Authors

American Museum of Natural History
New York, NY

Michelle Anderson, MS
Lecturer
The Ohio State University
Columbus, OH

Juli Berwald, PhD
Science Writer
Austin, TX

John F. Bolzan, PhD
Science Writer
Columbus, OH

Rachel Clark, MS
Science Writer
Moscow, ID

Patricia Craig, MS
Science Writer
Bozeman, MT

Randall Frost, PhD
Science Writer
Pleasanton, CA

Lisa S. Gardiner, PhD
Science Writer
Denver, CO

Jennifer Gonya, PhD
The Ohio State University
Columbus, OH

Mary Ann Grobbel, MD
Science Writer
Grand Rapids, MI

Whitney Crispen Hagins, MA, MAT
Biology Teacher
Lexington High School
Lexington, MA

Carole Holmberg, BS
Planetarium Director
Calusa Nature Center and Planetarium, Inc.
Fort Myers, FL

Tina C. Hopper
Science Writer
Rockwall, TX

Jonathan D. W. Kahl, PhD
Professor of Atmospheric Science
University of Wisconsin-Milwaukee
Milwaukee, WI

Nanette Kalis
Science Writer
Athens, OH

S. Page Keeley, MEd
Maine Mathematics and Science Alliance
Augusta, ME

Cindy Klevickis, PhD
Professor of Integrated Science and Technology
James Madison University
Harrisonburg, VA

Kimberly Fekany Lee, PhD
Science Writer
La Grange, IL

Michael Manga, PhD
Professor
University of California, Berkeley
Berkeley, CA

Devi Ried Mathieu
Science Writer
Sebastopol, CA

Elizabeth A. Nagy-Shadman, PhD
Geology Professor
Pasadena City College
Pasadena, CA

William D. Rogers, DA
Professor of Biology
Ball State University
Muncie, IN

Donna L. Ross, PhD
Associate Professor
San Diego State University
San Diego, CA

Marion B. Sewer, PhD
Assistant Professor
School of Biology
Georgia Institute of Technology
Atlanta, GA

Julia Meyer Sheets, PhD
Lecturer
School of Earth Sciences
The Ohio State University
Columbus, OH

Michael J. Singer, PhD
Professor of Soil Science
Department of Land, Air and Water Resources
University of California
Davis, CA

Karen S. Sottosanti, MA
Science Writer
Pickerington, Ohio

Paul K. Strode, PhD
I.B. Biology Teacher
Fairview High School
Boulder, CO

Jan M. Vermilye, PhD
Research Geologist
Seismo-Tectonic Reservoir Monitoring (STRM)
Boulder, CO

Judith A. Yero, MA
Director
Teacher's Mind Resources
Hamilton, MT

Dinah Zike, MEd
Author, Consultant,
Inventor of Foldables
Dinah Zike Academy;
Dinah-Might Adventures, LP
San Antonio, TX

Margaret Zorn, MS
Science Writer
Yorktown, VA

Consulting Authors

Alton L. Biggs
Biggs Educational Consulting
Commerce, TX

Ralph M. Feather, Jr., PhD
Assistant Professor
Department of Educational
Studies and Secondary
Education
Bloomsburg University
Bloomsburg, PA

Douglas Fisher, PhD
Professor of Teacher Education
San Diego State University
San Diego, CA

Edward P. Ortleb
Science/Safety Consultant
St. Louis, MO

Series Consultants

Science

Solomon Bililign, PhD
Professor
Department of Physics
North Carolina Agricultural
and Technical State University
Greensboro, NC

John Choinski
Professor
Department of Biology
University of Central Arkansas
Conway, AR

Anastasia Chopelas, PhD
Research Professor
Department of Earth and
Space Sciences
UCLA
Los Angeles, CA

David T. Crowther, PhD
Professor of Science Education
University of Nevada, Reno
Reno, NV

A. John Gatz
Professor of Zoology
Ohio Wesleyan University
Delaware, OH

Sarah Gille, PhD
Professor
University of California
San Diego
La Jolla, CA

David G. Haase, PhD
Professor of Physics
North Carolina State
University
Raleigh, NC

Janet S. Herman, PhD
Professor
Department of Environmental
Sciences
University of Virginia
Charlottesville, VA

David T. Ho, PhD
Associate Professor
Department of Oceanography
University of Hawaii
Honolulu, HI

Ruth Howes, PhD
Professor of Physics
Marquette University
Milwaukee, WI

Jose Miguel Hurtado, Jr., PhD
Associate Professor
Department of Geological
Sciences
University of Texas at El Paso
El Paso, TX

Monika Kress, PhD
Assistant Professor
San Jose State University
San Jose, CA

Mark E. Lee, PhD
Associate Chair & Assistant
Professor
Department of Biology
Spelman College
Atlanta, GA

Linda Lundgren
Science writer
Lakewood, CO

Series Consultants, continued

Keith O. Mann, PhD
Ohio Wesleyan University
Delaware, OH

Charles W. McLaughlin, PhD
Adjunct Professor of Chemistry
Montana State University
Bozeman, MT

Katharina Pahnke, PhD
Research Professor
Department of Geology and Geophysics
University of Hawaii
Honolulu, HI

Jesús Pando, PhD
Associate Professor
DePaul University
Chicago, IL

Hay-Oak Park, PhD
Associate Professor
Department of Molecular Genetics
Ohio State University
Columbus, OH

David A. Rubin, PhD
Associate Professor of Physiology
School of Biological Sciences
Illinois State University
Normal, IL

Toni D. Sauncy
Assistant Professor of Physics
Department of Physics
Angelo State University
San Angelo, TX

Malathi Srivatsan, PhD
Associate Professor of Neurobiology
College of Sciences and Mathematics
Arkansas State University
Jonesboro, AR

Cheryl Wistrom, PhD
Associate Professor of Chemistry
Saint Joseph's College
Rensselaer, IN

Reading

ReLeah Cossett Lent
Author/Educational Consultant
Blue Ridge, GA

Math

Vik Hovsepian
Professor of Mathematics
Rio Hondo College
Whittier, CA

Series Reviewers

Thad Boggs
Mandarin High School
Jacksonville, FL

Catherine Butcher
Webster Junior High School
Minden, LA

Erin Darichuk
West Frederick Middle School
Frederick, MD

Joanne Hedrick Davis
Murphy High School
Murphy, NC

Anthony J. DiSipio, Jr.
Octorara Middle School
Atglen, PA

Adrienne Elder
Tulsa Public Schools
Tulsa, OK

Series Reviewers, continued

Carolyn Elliott
Iredell-Statesville Schools
Statesville, NC

Christine M. Jacobs
Ranger Middle School
Murphy, NC

Jason O. L. Johnson
Thurmont Middle School
Thurmont, MD

Felecia Joiner
Stony Point Ninth Grade Center
Round Rock, TX

Joseph L. Kowalski, MS
Lamar Academy
McAllen, TX

Brian McClain
Amos P. Godby High School
Tallahassee, FL

Von W. Mosser
Thurmont Middle School
Thurmont, MD

Ashlea Peterson
Heritage Intermediate Grade Center
Coweta, OK

Nicole Lenihan Rhoades
Walkersville Middle School
Walkersvillle, MD

Maria A. Rozenberg
Indian Ridge Middle School
Davie, FL

Barb Seymour
Westridge Middle School
Overland Park, KS

Ginger Shirley
Our Lady of Providence Junior-Senior High School
Clarksville, IN

Curtis Smith
Elmwood Middle School
Rogers, AR

Sheila Smith
Jackson Public School
Jackson, MS

Sabra Soileau
Moss Bluff Middle School
Lake Charles, LA

Tony Spoores
Switzerland County Middle School
Vevay, IN

Nancy A. Stearns
Switzerland County Middle School
Vevay, IN

Kari Vogel
Princeton Middle School
Princeton, MN

Alison Welch
Wm. D. Slider Middle School
El Paso, TX

Linda Workman
Parkway Northeast Middle School
Creve Coeur, MO

Teacher Advisory Board

The Teacher Advisory Board gave the authors, editorial staff, and design team feedback on the content and design of the Student Edition. They provided valuable input in the development of *Glencoe ⓘScience*.

Frances J. Baldridge
Department Chair
Ferguson Middle School
Beavercreek, OH

Jane E. M. Buckingham
Teacher
Crispus Attucks Medical
Magnet High School
Indianapolis, IN

Elizabeth Falls
Teacher
Blalack Middle School
Carrollton, TX

Nelson Farrier
Teacher
Hamlin Middle School
Springfield, OR

Michelle R. Foster
Department Chair
Wayland Union
Middle School
Wayland, MI

Rebecca Goodell
Teacher
Reedy Creek Middle School
Cary, NC

Mary Gromko
Science Supervisor K–12
Colorado Springs District 11
Colorado Springs, CO

Randy Mousley
Department Chair
Dean Ray Stucky
Middle School
Wichita, KS

David Rodriguez
Teacher
Swift Creek Middle School
Tallahassee, FL

Derek Shook
Teacher
Floyd Middle Magnet School
Montgomery, AL

Karen Stratton
Science Coordinator
Lexington School District One
Lexington, SC

Stephanie Wood
Science Curriculum Specialist,
K–12
Granite School District
Salt Lake City, UT

Online Guide

Get ConnectED
connectED.mcgraw-hill.com

ConnectED
▶ Your Digital Science Portal

 Video
See the science in real life through these exciting videos.

 **Audio**
Click the link and you can listen to the text while you follow along.

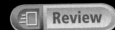 **Review**
Try these interactive tools to help you review the lesson concepts.

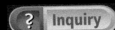 **Inquiry**
Explore concepts through hands-on and virtual labs.

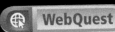 **WebQuest**
These web-based challenges relate the concepts you're learning about to the latest news and research.

Digital and Print Solutions

The icons in your online student edition link you to interactive learning opportunities. Browse your online student book to find more.

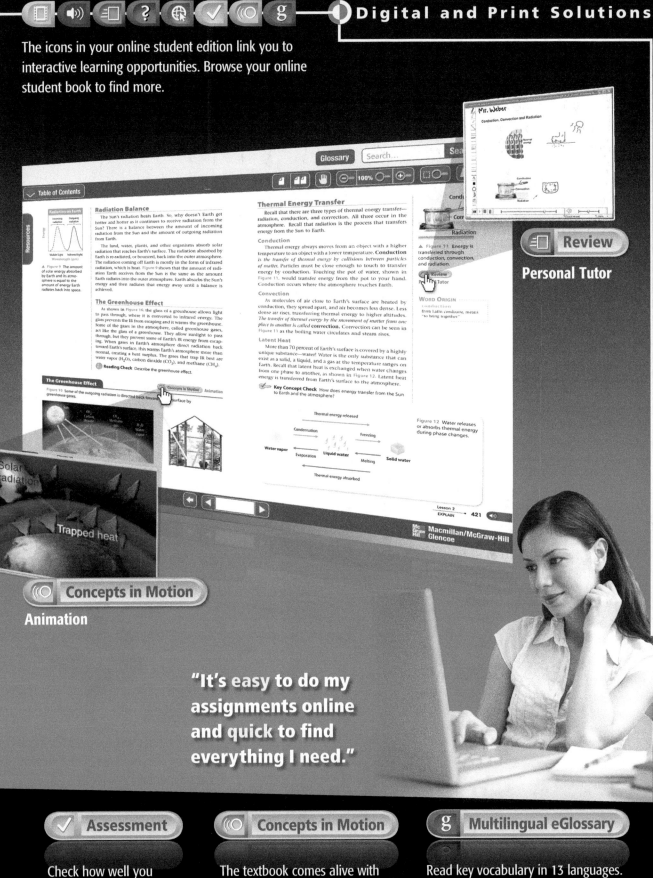

Review — Personal Tutor

Concepts in Motion — Animation

"It's easy to do my assignments online and quick to find everything I need."

Assessment
Check how well you understand the concepts with online...

Concepts in Motion
The textbook comes alive with animated explanations of important concepts...

Multilingual eGlossary
Read key vocabulary in 13 languages.

Treasure Hunt

Your science book has many features that will aid you in your learning. Some of these features are listed below. You can use the activity at the right to help you find these and other special features in the book.

- **THE BIG IDEA** can be found at the start of each chapter.
- The Reading Guide at the start of each lesson lists **Key Concepts**, vocabulary terms, and online supplements to the content.
- **ConnectED** icons direct you to online resources such as animations, personal tutors, math practices, and quizzes.
- **Inquiry** Labs and Skill Practices are in each chapter.
- Your **FOLDABLES** help organize your notes.

START

1. What four margin items can help you build your vocabulary?

2. On what page does the glossary begin? What glossary is online?

3. In which Student Resource at the back of your book can you find a listing of Laboratory Safety Symbols?

4. Suppose you want to find a list of all the Launch Labs, MiniLabs, Skill Practices, and Labs, where do you look?

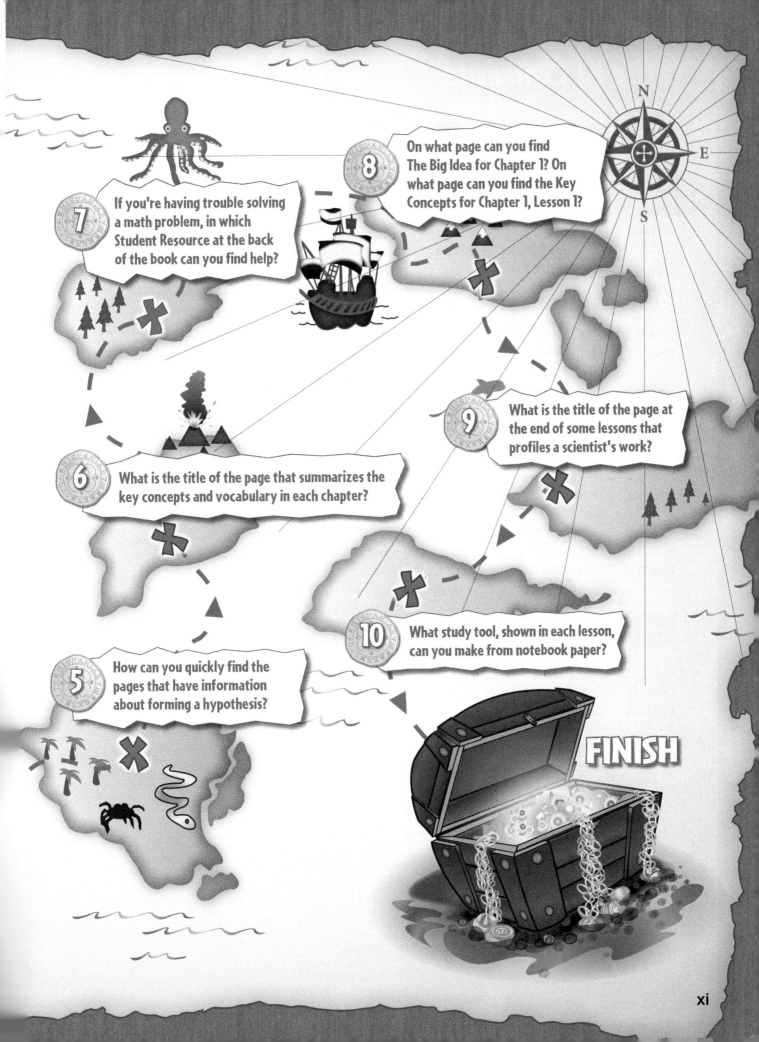

Table of Contents

Unit 5 — Exploring the Universe ... 682

Chapter 19 Exploring Space ... 686
- Lesson 1 Observing the Universe ... 688
  - **Skill Practice** How can you construct a simple telescope? ... 697
- Lesson 2 Early History of Space Exploration ... 698
- Lesson 3 Recent and Future Space Missions ... 706
  - **Lab** Design and Construct a Moon Habitat ... 714

Chapter 20 The Sun-Earth-Moon System ... 722
- Lesson 1 Earth's Motion ... 724
  - **Skill Practice** How does Earth's tilted rotation axis affect the seasons? ... 733
- Lesson 2 Earth's Moon ... 734
- Lesson 3 Eclipses and Tides ... 742
  - **Lab** Phases of the Moon ... 750

Chapter 21 The Solar System ... 758
- Lesson 1 The Structure of the Solar System ... 760
- Lesson 2 The Inner Planets ... 768
  - **Skill Practice** What can we learn about planets by graphing their characteristics? ... 775
- Lesson 3 The Outer Planets ... 776
- Lesson 4 Dwarf Planets and Other Objects ... 784
  - **Lab** Scaling down the Solar System ... 790

Chapter 22 Stars and Galaxies ... 798
- Lesson 1 The View from Earth ... 800
  - **Skill Practice** How can you use scientific illustrations to locate constellations? ... 807
- Lesson 2 The Sun and Other Stars ... 808
- Lesson 3 Evolution of Stars ... 816
  - **Skill Practice** How can graphing data help you understand stars? ... 823
- Lesson 4 Galaxies and the Universe ... 824
  - **Lab** Describe a Trip Through Space ... 832

Table of Contents

Student Resources

Science Skill Handbook **SR-2**
   Scientific Methods SR-2
   Safety Symbols SR-11
   Safety in the Science Laboratory SR-12

Math Skill Handbook **SR-14**
   Math Review SR-14
   Science Application SR-24

Foldables Handbook **SR-29**

Reference Handbook **SR-40**
   Periodic Table of the Elements SR-40
   Topographic Map Symbols SR-42
   Rocks SR-43
   Minerals SR-44
   Weather Map Symbols SR-46

Glossary **G-2**

Index **I-2**

Credits **C-2**

Inquiry

Launch Labs

19-1	Do you see what I see?	689
19-2	How do rockets work?	699
19-3	How is gravity used to send spacecraft farther in space?	707
20-1	Does Earth's shape affect temperatures on Earth's surface?	725
20-2	Why does the Moon appear to change shape?	735
20-3	How do shadows change?	743
21-1	How do you know which distance unit to use?	761
21-2	What affects the temperature on the inner planets?	769
21-3	How do we see distant objects in the solar system?	777
21-4	How might asteroids and moons form?	785
22-1	How can you "see" invisible energy?	801
22-2	What are those spots on the Sun?	809
22-3	Do stars have life cycles?	817
22-4	Does the universe move?	825

MiniLabs

19-1	What is white light?	691
19-2	How does lack of friction in space affect simple tasks?	703
19-3	What conditions are required for life on Earth?	710
20-1	What keeps Earth in orbit?	726
20-2	How can the Moon be rotating if the same side of the Moon is always facing Earth?	738
20-3	What does the Moon's shadow look like?	744
21-1	How can you model an elliptical orbit?	765
21-2	How can you model the inner planets?	772
21-3	How do Saturn's moons affect its rings?	781
21-4	How do impact craters form?	788
22-1	How does light differ?	803
22-2	Can you model the Sun's structure?	810
22-3	How do astronomers detect black holes?	820
22-4	Can you identify a galaxy?	827

Inquiry

Inquiry Skill Practice

19-1 How can you construct a simple telescope? 697
20-1 How does Earth's tilted rotation axis affect the seasons? 733
21-2 What can we learn about planets by graphing their characteristics? 775
22-1 How can you use scientific illustrations to locate constellations? 807
22-3 How can graphing data help you understand stars? 823

Inquiry Labs

19-3 Design and Construct a Moon Habitat 714
20-3 Phases of the Moon 750
21-4 Scaling down the Solar System 790
22-4 Describe a Trip Through Space 832

Features

HOW IT WORKS

19-2 Going Up 705
22-2 Viewing the Sun in 3-D 815

SCIENCE & SOCIETY

20-2 Return to the Moon 741

CAREERS in SCIENCE

21-3 Pluto 783

Unit 5

EXPLORING THE UNIVERSE

"IF YOU LOOK TO YOUR RIGHT, YOU'LL SEE OUR CLOSEST NEIGHBOR, THE ANDROMEDA GALAXY…"

"ONLY 2.5 MILLION LIGHT-YEARS AWAY."

"NEXT, WE HAVE A DYING STAR THAT HAS EXPANDED INTO A RED GIANT."

2000 B.C. — **1600** — **1700** — **1800**

1600 B.C. Babylonian texts show records of people observing Venus without the aid of technology. Its appearance is recorded for 21 years.

265 B.C. Greek astronomer Timocharis makes the first recorded observation of Mercury.

1610 Galileo Galilei observes the four largest moons of Jupiter through his telescope.

1613 Galileo records observations of the planet Neptune but mistakes it for a star.

1655 Astronomer Christiaan Huygens observes Saturn and discovers its rings, which were previously thought to be large moons on each side.

1781 William Herschel discovers the planet Uranus.

Unit 5 Nature of SCIENCE

Technology

It may sound strange, but some of the greatest benefits of the space program are benefits to life here on Earth. Devices ranging from hand-held computers to electric socks rely on technologies first developed for space exploration. **Technology** is the practical application of science to commerce or industry. Space technologies have increased our understanding of Earth and our ability to locate and conserve resources.

Problems, such as how best to explore the solar system and outer space, often send scientists on searches for new knowledge. Engineers use scientific knowledge to develop new technologies for space. Then, some of those technologies are modified to solve problems on Earth. For example, lightweight solar panels on the outside of a spacecraft convert the Sun's energy into electricity that powers the spacecraft for long space voyages. Similar but smaller, flexible solar panels, as shown in **Figure 1** are now available for consumers to purchase. They can be used to power small electronics when traveling. **Figure 2** shows how other technologies from space help conserve natural resources.

Figure 1 Lightweight, flexible solar cells developed for spacecraft help to conserve Earth's resources.

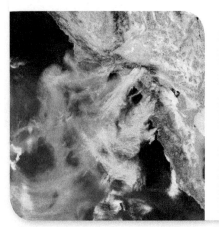

This image was taken by the *Terra* satellite and shows fires burning in California. The image helps firefighters see the size and the location of the fires. It also helps scientists study the effect of fires on Earth's atmosphere. ▼

Some portable water purification kits use technologies developed to provide safe, clean drinking water for astronauts. This kit can provide clean, safe drinking water for an entire village in a remote area or supply drinking water after a natural disaster.

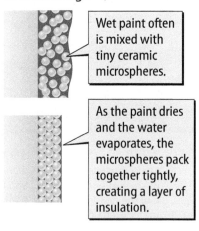

Engineers developed glass spheres about the size of a grain of flour to insulate super-cold spacecraft fuel lines. Similar microspheres act as insulators when mixed with paints. This technology can help reduce the energy needed to heat and cool buildings. ▼

- Wet paint often is mixed with tiny ceramic microspheres.
- As the paint dries and the water evaporates, the microspheres pack together tightly, creating a layer of insulation.

Figure 2 Some technologies developed as part of the space program have greatly benefited life on Earth.

Figure 3 The satellite image on the left is similar to what you would see with your eyes from space. A satellite sensor that detects other wavelengths of light produced the colored satellite image on the right. It shows the locations of nearly a dozen different minerals.

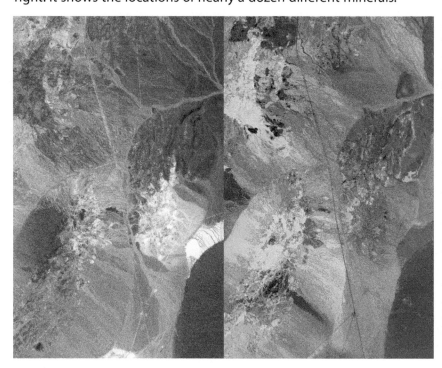

Inquiry MiniLab
25 minutes

How would you use space technology?

Many uses for space technology haven't been discovered yet. Can you develop one?

1. Identify a problem locating, protecting, or preserving resources that could be solved using space technology.
2. Prepare a short oral presentation explaining your technology.

Analyze and Conclude

1. **Describe** How is your technology used in space?
2. **Explain** How do you use technology to solve a problem on Earth?

Solving Problems and Improving Abilities

Science and technology depend on each other. For example, images from space greatly improve our understanding of Earth. **Figure 3** above shows a satellite image of a Nevada mine. The satellite is equipped with sensors that detect visible light, much like your eyes do. The image on the right shows a satellite image of the same site taken with a sensor that detects wavelengths of light your eyes cannot see. This image provides information about the types of minerals in the mine. Each color in the image on the right shows the location of a different mineral, reducing the time it takes geologists to locate mineral deposits.

Scientists use other kinds of satellite sensors for different purposes. Engineers have modified space technology to produce satellite images of cloud cover over Earth's surface, as shown in **Figure 4**. Images like this one improve global weather forecasting and help scientists understand changes in Earth's atmosphere. Of course, science can answer only some of society's questions, and technology cannot solve all problems. But together, they can improve the quality of life for all.

Figure 4 This satellite image shows reflection of sunlight (yellow), deep clouds (white), low clouds (pale yellow), high clouds (blue), vegetation (green), and sea (dark).

Technology • 685

Chapter 19

Exploring Space

 How do humans observe and explore space?

Inquiry Can satellites see into space?

Yes, they can! The satellite shown here is a telescope. It collects light from distant objects in space. But, most satellites you might be familiar with point toward Earth. They provide navigation assistance, monitor weather, and bounce communication signals to and from Earth.

- Why would scientists want to put a telescope in space?
- In what other ways do scientists observe and explore space?
- What are goals of some current and future space missions?

Get Ready to Read

What do you think?
Before you read, decide if you agree or disagree with each of these statements. As you read this chapter, see if you change your mind about any of the statements.

1. Astronomers put telescopes in space to be closer to the stars.
2. Telescopes can work only using visible light.
3. Humans have walked on the Moon.
4. Some orthodontic braces were developed using space technology.
5. Humans have landed on Mars.
6. Scientists have detected water on other bodies in the solar system.

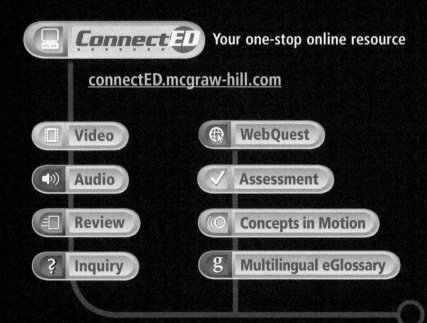

ConnectED Your one-stop online resource

connectED.mcgraw-hill.com

- Video
- WebQuest
- Audio
- Assessment
- Review
- Concepts in Motion
- Inquiry
- Multilingual eGlossary

Lesson 1

Reading Guide

Key Concepts
ESSENTIAL QUESTIONS

- How do scientists use the electromagnetic spectrum to study the universe?
- What types of telescopes and technology are used to explore space?

Vocabulary

electromagnetic spectrum p. 690

refracting telescope p. 692

reflecting telescope p. 692

radio telescope p. 693

Multilingual eGlossary

Video   Science Video

Observing the Universe

Inquiry How can you see this?

This is an expanding halo of dust in space, illuminated by the light from the star in the center. This photo was taken with a telescope. How do you think telescopes obtain such clear images?

Inquiry Launch Lab

15 minutes

Do you see what I see?

Your eyes have lenses. Eyeglasses, cameras, telescopes, and many other tools involving light also have lenses. Lenses are transparent materials that refract light, or cause light to change direction. Lenses can cause light rays to form images as they come together or move apart.

1. Read and complete a lab safety form.
2. Place each of the **lenses** on the words of this sentence.
3. Slowly move each lens up and down over the words to observe if or how the words change. Record your observations in your Science Journal.
4. Hold each lens at arm's length and focus on an object a few meters away. Observe how the object looks through each lens. Make simple drawings to illustrate what you observe.

Think About This

1. What happened to the words as you moved the lenses toward and away from the sentence?
2. What did the distant object look like through each lens?
3. **Key Concept** How do you think lenses are used in telescopes to explore space?

Observing the Sky

If you look up at the sky on a clear night, you might be able to see the Moon, planets, and stars. These objects have not changed much since people first turned their gaze skyward. People in the past spent a lot of time observing the sky. They told stories about the stars, and they used stars to tell time. Most people thought Earth was the center of the universe.

Astronomers today know that Earth is part of a system of eight planets revolving around the Sun. The Sun, in turn, is part of a larger system called the Milky Way galaxy that contains billions of other stars. And the Milky Way is one of billions of other galaxies in the universe. As small as Earth might seem in the universe, it could be unique. Scientists have not found life anywhere else.

One advantage astronomers have over people in the past is the *telescope*. Telescopes enable astronomers to observe many more stars than they could with their eyes alone. Telescopes gather and focus light from objects in space. The photo on the opposite page was taken with a telescope that orbits Earth. Astronomers use many kinds of telescopes to study the energy emitted by stars and other objects in space.

Reading Check What is the purpose of telescopes?

WORD ORIGIN

telescope
from Greek *tele*, means "far"; and Greek *skopos*, means "seeing"

Electromagnetic Waves

Stars emit energy that radiates into space as electromagnetic (ih lek troh mag NEH tik) waves. Electromagnetic waves are different from mechanical waves, such as sound waves. Sound waves can transfer energy through solids, liquids, or gases. Electromagnetic waves can transfer energy through matter or through a vacuum, such as space. The energy they carry is called radiant energy.

The Electromagnetic Spectrum

The entire range of radiant energy carried by electromagnetic waves is the **electromagnetic spectrum.** As shown in **Figure 1**, waves of the electromagnetic spectrum are continuous. They range from gamma rays with short wavelengths at one end to radio waves with long wavelengths at the other end. Radio waves can be thousands of kilometers in length. Gamma rays can be smaller in length than the size of an atom.

Reading Check How is radiant energy carried in space?

Humans observe only a small part of the electromagnetic spectrum—the visible part in the middle. Visible light includes all the colors you see. You cannot see the other parts of the electromagnetic spectrum, but you can use them. When you talk on a cellular phone, you use microwaves. When you change the TV channel with a remote-control device, you use infrared waves.

Radiant Energy and Stars

Most stars emit energy in all wavelengths. But how much of each wavelength they emit depends on their temperatures. Hot stars emit mostly shorter waves with higher energy, such as X-rays, gamma rays, and ultraviolet waves. Cool stars emit mostly longer waves with lower energy, such as infrared waves and radio waves. The Sun has a medium temperature range. It emits much of its energy as visible light.

FOLDABLES

Use seven quarter-sheets of paper to make a tabbed diagram illustrating the electromagnetic spectrum. In the middle section of a large shutterfold project, tape or glue the left edges of the tabs so they overlap to illustrate the varying sizes of the waves from longest to shortest.

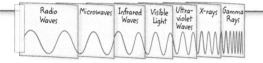

Figure 1 Objects emit radiation in continuous wavelengths. Most wavelengths are not visible to the human eye.

Visual Check Approximately how long are the wavelengths of microwaves?

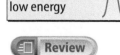

Inquiry MiniLab

15 minutes

What is white light?

Sunlight and the light from an ordinary lightbulb are both examples of visible light. You might think that white light is all white. Is it?

1. Read and complete a lab safety form.
2. Darken the room, and shine a **flashlight** through a **prism** on a flat surface. Adjust the positions of the prism and the flashlight until you observe the entire visible light spectrum.
3. In your Science Journal, use **colored pencils** to draw what you see.

Analyze and Conclude

1. **Define** What is white light?
2. **Compare and Contrast** Which component of white light has the longest wavelength? Which has the shortest wavelength? Explain your answers.
3. **Key Concept** How does visible light fit into the electromagnetic spectrum?

Why You See Planets and Moons

Planets and moons are much cooler than even the coolest stars. They do not make their own energy and, therefore, do not emit light. However, you can see the Moon and the planets because they reflect light from the Sun.

Light from the Past

All electromagnetic waves, from radio waves to gamma rays, travel through space at a constant speed of 300,000 km/s. This is called the speed of light. The speed of light might seem incredibly fast, but the universe is very large. Even moving at the speed of light, it can take millions or billions of years for some light waves to reach Earth because of the large distances in space.

Because it takes time for light to travel, you see planets and stars as they were when their light started its journey to Earth. It takes very little time for light to travel within the solar system. Reflected light from the Moon reaches Earth in about 1 second. Light from the Sun reaches Earth in about 8 minutes. It reaches Jupiter in about 40 minutes.

Light from stars is much older. Some stars are so far away that it can take millions or billions of years for their radiant energy to reach Earth. Therefore, by studying energy from stars, astronomers can learn what the universe was like millions or billions of years ago.

 Reading Check How is looking at stars like looking at the past?

Math Skills

Scientific Notation

Scientists use scientific notation to work with large numbers. Express the speed of light in scientific notation using the following process.

1. Move the decimal point until only one nonzero digit remains on the left.

   300,000 → 3.00000

2. Use the number of places the decimal point moved (5) as a power of ten.

   300,000 km/s = 3.0×10^5 km/s

Practice

The Sun is 150,000,000 km from Earth. Express this distance in scientific notation.

- Math Practice
- Personal Tutor

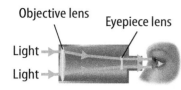

Refracting telescope

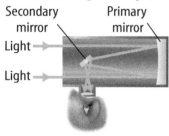

Reflecting telescope

▲ **Figure 2** Optical telescopes collect visible light in two different ways.

Earth-Based Telescopes

Telescopes are designed to collect a certain type of electromagnetic wave. Some telescopes detect visible light, and others detect radio waves and microwaves.

Optical Telescopes

There are two kinds of optical telescopes—refracting telescopes and reflecting telescopes, illustrated in **Figure 2.**

Refracting Telescopes Have you ever used a magnifying lens? You might have noticed that the lens was curved and thick in the middle. This is a convex lens. *A telescope that uses a convex lens to concentrate light from a distant object is a* **refracting telescope.** As shown at the top of **Figure 2,** the objective lens in a refracting telescope is the lens closest to the object being viewed. The light goes through the objective lens and refracts, forming a small, bright image. The eyepiece is a second lens that magnifies the image.

 Key Concept Check Which electromagnetic waves do refracting telescopes collect?

Reflecting Telescopes Most large telescopes use curved mirrors instead of curved lenses. *A telescope that uses a curved mirror to concentrate light from a distant object is a* **reflecting telescope.** As shown at the bottom of **Figure 2,** light is reflected from a primary mirror to a secondary mirror. The secondary mirror is tilted to allow the viewer to see the image. Generally, larger primary mirrors produce clearer images than smaller mirrors. However, there is a limit to mirror size. The largest reflecting telescopes, such as the Keck Telescopes on Hawaii's Mauna Kea, shown in **Figure 3,** have many small mirrors linked together. These small mirrors act as one large primary mirror.

Figure 3 Each 10-m primary mirror in the twin Keck Telescopes consists of 36 small mirrors. ▼

Radio Telescope

▲ Figure 4 Radio telescopes are often built in large arrays. Computers convert radio data into images.

Radio Telescopes

Unlike a telescope that collects visible light waves, a **radio telescope** *collects radio waves and some microwaves using an antenna that looks like a TV satellite dish.* Because these waves have long wavelengths and carry little energy, radio antennae must be large to collect them. Radio telescopes are often built together and used as if they were one telescope. The telescopes shown in **Figure 4** are part of the Very Large Array in New Mexico. The 27 instruments in this array act as a single telescope with a 36-km diameter.

 Reading Check Why are radio telescopes built together in large arrays?

Distortion and Interference

Moisture in Earth's atmosphere can absorb and distort radio waves. Therefore, most radio telescopes are located in remote deserts, which have dry environments. Remote deserts also tend to be far from radio stations, which emit radio waves that interfere with radio waves from space.

Water vapor and other gases in Earth's atmosphere also distort visible light. Stars seem to twinkle because gases in the atmosphere move, refracting the light. This causes the location of a star's image to change slightly. At high elevations, the atmosphere is thin and produces less distortion than it does at low elevations. That is why most optical telescopes are built on mountains. New technology called adaptive optics lessens the effects of atmospheric distortion even more, as shown in **Figure 5**.

Figure 5 Adaptive optics sharpens images by reducing atmospheric distortion. ▼

Before Adaptive Optics

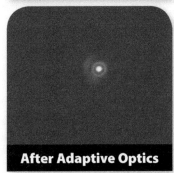
After Adaptive Optics

Lesson 1
EXPLAIN

Wavelengths from Space

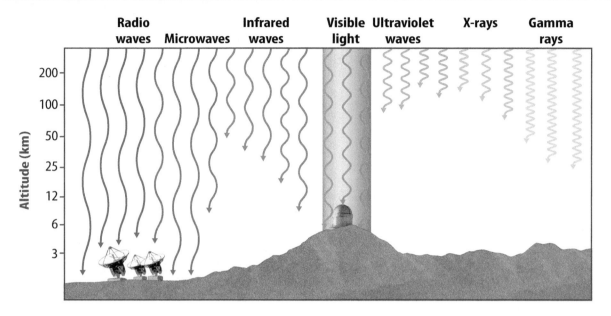

▲ **Figure 6** Most electromagnetic waves do not penetrate Earth's atmosphere. Even though the atmosphere blocks most UV rays, some still reach Earth's surface.

Visual Check About how far above Earth's surface do gamma waves reach?

Figure 7 The *Hubble Space Telescope* is controlled by astronomers on Earth. ▼

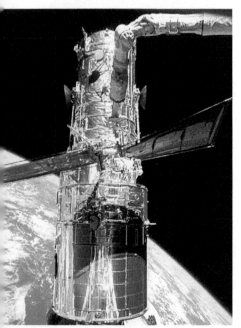

Space Telescopes

Why would astronomers want to put a telescope in space? The reason is Earth's atmosphere. Earth's atmosphere absorbs some types of electromagnetic radiation. As shown in **Figure 6,** visible light, radio waves, and some microwaves reach Earth's surface. But other types of electromagnetic waves do not. Telescopes on Earth can collect only the electromagnetic waves that are not absorbed by Earth's atmosphere. Telescopes in space can collect energy at all wavelengths, including those that Earth's atmosphere would absorb, such as most infrared light, most ultraviolet light, and X-rays.

 Key Concept Check Why do astronomers put some telescopes in space?

Optical Space Telescopes

Optical telescopes collect visible light on Earth's surface, but optical telescopes work better in space. The reason, again, is Earth's atmosphere. As you read earlier, gases in the atmosphere can absorb some wavelengths. In space, there are no atmospheric gases. The sky is darker, and there is no weather.

The first optical space telescope was launched in 1990. The *Hubble Space Telescope,* shown in **Figure 7,** is a reflecting telescope that orbits Earth. Its primary mirror is 2.4 m in diameter. At first the *Hubble* images were blurred because of a flaw in the mirror. In 1993, astronauts repaired the telescope. Since then, *Hubble* has routinely sent to Earth spectacular images of far-distant objects. The photo at the beginning of this lesson was taken with the *Hubble* telescope.

Using Other Wavelengths

The *Hubble Space Telescope* is the only space telescope that collects visible light. Dozens of other space telescopes, operated by many different countries, gather ultraviolet, X-ray, gamma ray, and infrared light. Each type of telescope can point at the same region of sky and produce a different image. The image of the star Cassiopeia A (ka see uh PEE uh • AY) in **Figure 8** was made with a combination of optical, X-ray, and infrared data. The colors represent different kinds of material left over from the star's explosion many years ago.

Spitzer Space Telescope Young stars and planets hidden by dust and gas cannot be viewed in visible light. However, infrared wavelengths can penetrate the dust and reveal what is beyond it. Infrared can also be used to observe objects too old and too cold to emit visible light. In 2003, the *Spitzer Space Telescope* was launched to collect infrared waves, as it orbits the Sun.

 Reading Check Which type of radiant energy does the *Spitzer Space Telescope* collect?

James Webb Space Telescope A larger space telescope, scheduled for launch in 2014, is also designed to collect infrared radiation as it orbits the Sun. The *James Webb Space Telescope*, illustrated in **Figure 9**, will have a mirror with an area 50 times larger than *Spitzer*'s mirror and seven times larger than *Hubble*'s mirror. Astronomers plan to use the telescope to detect galaxies that formed early in the history of the universe.

▲ Figure 8 Each color in this image of Cassiopeia A is derived from a different wavelength—yellow: visible; pink/red: infrared; green and blue: X-ray.

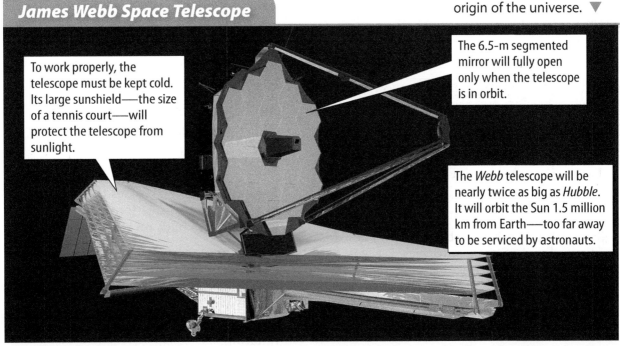

Figure 9 The advanced technology of the *James Webb Space Telescope* will help astronomers study the origin of the universe. ▼

James Webb Space Telescope

To work properly, the telescope must be kept cold. Its large sunshield—the size of a tennis court—will protect the telescope from sunlight.

The 6.5-m segmented mirror will fully open only when the telescope is in orbit.

The *Webb* telescope will be nearly twice as big as *Hubble*. It will orbit the Sun 1.5 million km from Earth—too far away to be serviced by astronauts.

Lesson 1 Review

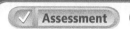

Visual Summary

Reflecting telescopes use mirrors to concentrate light.

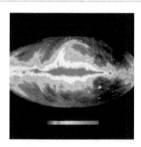
Earth-based telescopes can collect energy in the visible, radio, and microwave parts of the electromagnetic spectrum.

Space-based telescopes can collect wavelengths of energy that cannot penetrate Earth's atmosphere.

FOLDABLES

Use your lesson Foldable to review the lesson. Save your Foldable for the project at the end of the chapter.

What do you think NOW?

You first read the statements below at the beginning of the chapter.

1. Astronomers put telescopes in space to be closer to the stars.

2. Telescopes can work only using visible light.

Did you change your mind about whether you agree or disagree with the statements? Rewrite any false statements to make them true.

Use Vocabulary

1. **Distinguish** between a reflecting telescope and a refracting telescope.
2. **Use the term** *electromagnetic spectrum* in a sentence.
3. **Define** *radio telescope* in your own words.

Understand Key Concepts

4. Which emits visible light?
   - A. moon
   - B. planet
   - C. satellite
   - D. star
5. **Draw** a sketch that shows the difference in wavelength of a radio wave and a visible light wave. Which transfers more energy?
6. **Contrast** the *Hubble Space Telescope* and the *James Webb Space Telescope*.

Interpret Graphics

7. **Explain** The three images above represent the same area of sky. Explain why each looks different.
8. **Organize Information** Copy and fill in the graphic organizer below, listing the wavelengths collected by space telescopes, from the longest to the shortest.

Critical Thinking

9. **Suggest** a reason—besides the lessening of atmospheric distortion—why optical telescopes are built on remote mountains.

Math Skills

10. Light travels 9,460,000,000,000 km in 1 year. Express this number in scientific notation.

Inquiry: Follow a Procedure

30 minutes

How can you construct a simple telescope?

Have you ever looked at the night sky and wondered what you were looking at? Stars and planets look much the same. How can you distinguish them? In this lab, you will construct a simple telescope you can use to observe and distinguish distant objects.

Materials

lenses

cardboard tubes

silicon putty

wax pencil

rubber bands

masking tape

Safety

Learn It

In many science experiments, you must **follow a procedure** in order to know what materials to use and how to use them. In this activity, you will follow a procedure to construct a simple telescope.

Try It

1. Read and complete a lab safety form.

2. Move both lenses up and down over the print on this page to determine which lens has a shorter focal length. Use a marker to put a small dot on its edge. This will be your eyepiece lens.

3. Make a silicon putty rope 2–3 mm in diameter and about 15 cm long. Wrap the rope around the edge of one of the open ends of the smaller cardboard tube. Remove any extra putty.

4. Gently push the eyepiece lens onto the ring of putty. Wrap a piece of masking tape around the edge of the lens to secure it firmly.

5. Repeat steps 4 and 5 using the larger tube and the objective lens.

6. Place the smaller tube into the larger tube so that the eyepiece lens, in the smaller tube, extends outside the larger tube.

7. Use your telescope to view distant objects. Move the smaller tube in and out to focus your instrument. If possible, view the night sky with your telescope.
   ⚠ *Caution: Do not use your telescope or any other instrument to directly view the Sun.*

8. Record your observations in your Science Journal.

Apply It

9. **Identify** What type of telescope did you construct?

10. 🔑 **Key Concept** How does your telescope collect light?

Lesson 1 EXTEND • **697**

Lesson 2

Reading Guide

Key Concepts
ESSENTIAL QUESTIONS

- How are rockets and artificial satellites used?
- Why do scientists send both crewed and uncrewed missions into space?
- What are some ways that people use space technology to improve life on Earth?

Vocabulary
rocket p. 699
satellite p. 700
space probe p. 701
lunar p. 701
Project Apollo p. 702
space shuttle p. 702

 Multilingual eGlossary

 Video Science Video

Early History of Space Exploration

Inquiry Where is it headed?

Have you ever witnessed a rocket launch? Rockets produce gigantic clouds of smoke, long plumes of exhaust, and thundering noise. How are rockets used to explore space? What do they carry?

Inquiry Launch Lab

10 minutes

How do rockets work?

Space exploration would be impossible without rockets. Become a rocket scientist for a few minutes, and find out what sends rockets into space.

1. Read and complete a lab safety form.
2. Use **scissors** to carefully cut a 5-m piece of **string.**
3. Insert the string into a **drinking straw.** Tie each end of the string to a stationary object. Make sure the string is taut. Slide the drinking straw to one end of the string.
4. Blow up a **balloon.** Do not tie it. Instead, twist the neck and clamp it with a **clothespin** or a **paper clip. Tape** the balloon to the straw.
5. Remove the clothespin or paperclip to launch your rocket. Observe how the rocket moves. Record your observations in your Science Journal.

Think About This

1. Describe how your rocket moved along the string.
2. How might you get your rocket to go farther or faster?
3. **Key Concept** How do you think rockets are used in space exploration?

Rockets

Think about listening to a recording of your favorite music. Now think about how different it is to experience the same music at a live performance. This is like the difference between exploring space from a distance, with a telescope, and actually going there.

A big problem in launching an object into space is overcoming the force of Earth's gravity. This is accomplished with rockets. *A **rocket** is a vehicle designed to propel itself by ejecting exhaust gas from one end.* Fuel burned inside the rocket builds up pressure. The force from the exhaust thrusts the rocket forward, as shown in **Figure 10.** Rocket engines do not draw in oxygen from the surrounding air to burn their fuel, as jet engines do. They carry their oxygen with them. As a result, rockets can operate in space where there is very little oxygen.

Key Concept Check How are rockets used in space exploration?

Scientists launch rockets from Florida's Cape Canaveral Air Force Station or the Kennedy Space Center nearby. However, space missions are managed by scientists at several different research stations around the country.

Figure 10 Exhaust gases ejected from the end of a rocket push the rocket forward.

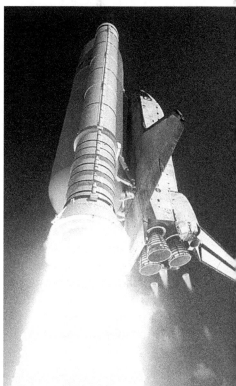

WORD ORIGIN
satellite
from Latin *satellitem*, means "attendant" or "bodyguard"

FOLDABLES
Make a vertical two-tab book. Record what you learn about crewed and uncrewed space missions under the tabs.

Artificial Satellites

Any small object that orbits a larger object is a **satellite**. The Moon is a natural satellite of Earth. Artificial satellites are made by people and launched by rockets. They orbit Earth or other bodies in space, transmitting radio signals back to Earth.

The First Satellites—*Sputnik* and *Explorer*

The first artificial, Earth-orbiting satellite was *Sputnik 1*. Many people think this satellite, launched in 1957 by the former Soviet Union, represents the beginning of the space age. In 1958, the United States launched its first Earth-orbiting satellite, *Explorer I*. Today, thousands of satellites orbit Earth.

How Satellites Are Used

The earliest satellites were developed by the military for navigation and to gather information. Today, Earth-orbiting satellites are also used to transmit television and telephone signals and to monitor weather and climate. An array of satellites called the Global Positioning System (GPS) is used for navigation in cars, boats, airplanes, and even for hiking.

Key Concept Check How are Earth-orbiting satellites used?

Early Exploration of the Solar System

In 1958, the U.S. Congress established the National Aeronautics and Space Administration (NASA). NASA oversees all U.S. space missions, including space telescopes. Some early steps in U.S. space exploration are shown in **Figure 11**.

Figure 11 Space exploration began with the first rocket launch in 1926.

Visual Check How many years after the first rocket was the first U.S. satellite launched into space?

Early Space Exploration

◀ **1926 First rocket:** Robert Goddard's liquid-fueled rocket rose 12 m into the air.

1958 First U.S. satellite: In the same year NASA was founded, *Explorer 1* was launched. It orbited Earth 58,000 times before burning up in Earth's atmosphere in 1970. ▶

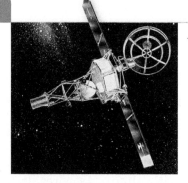

◀ **1962 First planetary probe:** *Mariner 2* traveled to Venus and collected data for 3 months. The craft now orbits the Sun.

1972 First probe to outer solar system: After flying past Jupiter, *Pioneer 10* is still traveling onward, someday to exit the solar system. ▶

Space Probes

Figure 12 Scientists use space probes to explore the planets and some moons in the solar system.

Visual Check Which type of probe might use a parachute?

Orbiter

Once orbiters reach their destinations, they use rockets to slow down enough to be captured in a planet's orbit. How long they orbit depends on their fuel supply. The orbiter probe here, *Pioneer,* orbited Venus.

Lander

Landers touch down on surfaces. Sometimes they release rovers. Landers use rockets and parachutes to slow their descent. The lander probe here, *Phoenix,* analyzed the Martian surface for evidence of water.

Flyby
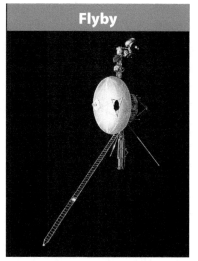
Flybys do not orbit or land. When its mission is complete, a flyby continues through space, eventually leaving the solar system. *Voyager 1,* here, explored Jupiter and Saturn and will soon leave the solar system.

Space Probes

Some spacecraft have human crews, but most do not. *A* **space probe** *is an uncrewed spacecraft sent from Earth to explore objects in space.* Space probes are robots that work automatically or by remote control. They take pictures and gather data. Probes are cheaper to build than crewed spacecraft, and they can make trips that would be too long or too dangerous for humans. Space probes are not designed to return to Earth. The data they gather are relayed to Earth via radio waves. **Figure 12** shows three major types of space probes.

Key Concept Check Why do scientists send uncrewed missions to space?

Lunar and Planetary Probes

The first probes to the Moon were sent by the United States and the former Soviet Union in 1959. Probes to the Moon are called lunar probes. *The term* **lunar** *refers to anything related to the Moon.* The first spacecraft to gather information from another planet was the flyby *Mariner 2,* sent to Venus in 1962. Since then, space probes have been sent to all the planets.

SCIENCE USE V. COMMON USE

probe

Science Use an uncrewed spacecraft

Common Use question or examine closely

Human Spaceflight

Sending humans into space was a major goal of the early space program. However, scientists worried about how radiation from the Sun and weightlessness in space might affect people's health. Because of this, they first sent dogs, monkeys, and chimpanzees. In 1961, the first human—an astronaut from the former Soviet Union—was launched into Earth's orbit. Shortly thereafter, the first American astronaut orbited Earth. Some highlights of the early U.S. human spaceflight program are shown in **Figure 13**.

The Apollo Program

In 1961, U.S. President John F. Kennedy challenged the American people to place a person on the Moon by the end of the decade. The result was **Project Apollo**—*a series of space missions designed to send people to the Moon.* In 1969, Neil Armstrong and Buzz Aldrin, Apollo 11 astronauts, were the first people to walk on the Moon.

Reading Check What was the goal of Project Apollo?

Space Transportation Systems

Early spacecraft and the rockets used to launch them were used only once. **Space shuttles** *are reusable spacecraft that transport people and materials to and from space.* Space shuttles return to Earth and land much like airplanes. NASA's fleet of space shuttles began operating in 1981. As the shuttles aged, NASA began developing a new transportation system, *Orion,* to replace them.

The *International Space Station*

The United States has its own space program. But it also cooperates with the space programs of other countries. In 1998, it joined 15 other nations to begin building the *International Space Station*. Occupied since 2000, this Earth-orbiting satellite is a research laboratory where astronauts from many countries work and live.

Research conducted aboard the *International Space Station* includes studying fungus, plant growth, and how human body systems react to low gravity conditions.

U.S. Human Spaceflight

Figure 13 Forty years after human spaceflight began, people were living and working in space.

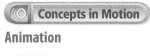

◀ A space shuttle piggybacked on rockets

▲ Apollo moon walk

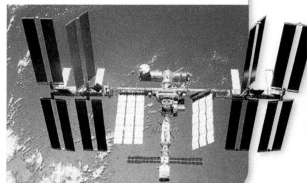
International Space Station orbiting Earth ▼

Inquiry MiniLab

15 minutes

How does lack of friction in space affect simple tasks?

Because objects are nearly weightless in space, there is little friction. What do you think might happen if an astronaut applied too much force when trying to move an object?

1. Read and complete a lab safety form.
2. Use **putty** to attach a **small thread spool** over the hole of a **CD.**
3. Inflate a **large, round balloon.** Twist the neck to keep the air inside. Stretch the neck of the balloon over the spool without releasing the air.
4. Place the CD on a smooth surface. Release the twist, and gently flick the CD with your finger. Describe your observations in your Science Journal.

Analyze and Conclude

1. **Infer** Why did the balloon craft move so easily?
2. **Draw Conclusions** How hard would it be to move a large object on the *International Space Station*?
3. **Key Concept** What challenges do astronauts face in space?

Space Technology

The space program requires materials that can withstand the extreme temperatures and pressures of space. Many of these materials have been applied to everyday life on Earth.

New Materials

Space materials must protect people from extreme conditions. They also must be flexible and strong. Materials developed for spacesuits are now used to make racing suits for swimmers, lightweight firefighting gear, running shoes, and other sports clothing.

Safety and Health

NASA developed a strong, fibrous material to make parachute cords for spacecraft that land on planets and moons. This material, five times stronger than steel, is used to make radial tires for automobiles.

Medical Applications

Artificial limbs, infrared ear thermometers, and robotic surgery all have roots in the space program. So do the orthodontic braces shown in **Figure 14.** These braces contain ceramic material originally developed to strengthen the heat resistance of space shuttles.

Key Concept Check What are some ways that space exploration has improved life on Earth?

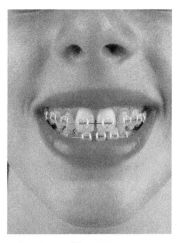

Figure 14 These braces contain a hard, strong ceramic originally developed for spacecraft.

Lesson 2
EXPLAIN

Lesson 2 Review

Visual Summary

Exhaust from burned fuel accelerates a rocket.

Some space probes can land on the surface of a planet or a moon.

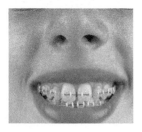

Technologies developed for the space program have been applied to everyday life on Earth.

FOLDABLES

Use your lesson Foldable to review the lesson. Save your Foldable for the project at the end of the chapter.

What do you think NOW?

You first read the statements below at the beginning of the chapter.

3. Humans have walked on the Moon.

4. Some orthodontic braces were developed using space technology.

Did you change your mind about whether you agree or disagree with the statements? Rewrite any false statements to make them true.

Use Vocabulary

1 Define *rocket* in your own words.

2 Use the term *satellite* in a sentence.

3 The mission that sent people to the Moon was _____.

Understand Key Concepts

4 What are rockets used for?
   A. carrying people
   B. launching satellites
   C. observing planets
   D. transmitting signals

5 Explain why *Sputnik 1* is considered the beginning of the space age.

6 Compare and contrast crewed and uncrewed space missions.

Interpret Graphics

7 Infer How is the balloon above like a rocket?

8 Organize Information Copy and fill in the graphic organizer below and use it to place the following in the correct order: *first human in space, invention of rockets, first human on the Moon, first artificial satellite.*

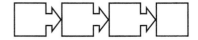

Critical Thinking

9 Predict how your life would be different if all artificial satellites stopped working.

10 Evaluate the benefits and drawbacks of international cooperation in space exploration.

Going Up

Could a space elevator make space travel easier?

HOW IT WORKS

If you wanted to travel into space, the first thing you would have to do is overcome the force of Earth's gravity. So far, the only way to do that has been to use rockets. Rockets are expensive, however. Many are used only once, and they require a lot of fuel. It takes a lot of resources to build and power a rocket. But what if you could take an elevator into space instead?

Space elevators were once science fiction, but scientists are now taking the possibility seriously. With the lightweight but strong materials under development today, experts say it could take only 10 years to build a space elevator. The image here shows how it might work.

It generally costs more than $100 million to place a 12,000-kg spacecraft into orbit using a rocket. Some people estimate that a space elevator could place the same craft into orbit for less than $120,000. A human passenger with luggage, together totaling 150 kg, might be able to ride the elevator to space for less than $1,500.

Counterweight: The spaceward end of the cable would attach to a captured asteroid or an artificial satellite. The asteroid or satellite would stabilize the cable and act as a counterweight.

Cable: Made of super-strong but thin materials, the cable would be the first part of the elevator to be built. A rocket-launched spacecraft would carry reels of cable into orbit. From there the cable would be unwound until one end reached Earth's surface.

Anchor Station: The cable's Earthward end would be attached here. A movable platform would allow operators to move the cable away from space debris in Earth's orbit that could collide with it. The platform would be movable because it would float on the ocean.

Climber: The "elevator car" would carry human passengers and objects into space. It could be powered by Earth-based laser beams, which would activate solar-cell "ears" on the outside of the car.

It's Your Turn

DEBATE Form an opinion about the space elevator and debate with a classmate. Could a space elevator become a reality in the near future? Would a space elevator benefit ordinary people? Should the space elevator be used for space tourism?

Lesson 3

Reading Guide

Key Concepts 🔑
ESSENTIAL QUESTIONS

- What are goals for future space exploration?
- What conditions are required for the existence of life on Earth?
- How can exploring space help scientists learn about Earth?

Vocabulary
extraterrestrial life p. 711
astrobiology p. 711

 Multilingual eGlossary

📹 Video

- Science Video
- What's Science Got to do With It?

Recent and Future Space Missions

Inquiry Blue Moon?

No, this is Mars! It is a false-color photo of an area on Mars where a future space probe might land. Scientists think the claylike material here might contain water and organic material. Could this material support life?

Inquiry Launch Lab

10 minutes

How is gravity used to send spacecraft farther in space?

Spacecraft use fuel to get to where they are going. But fuel is expensive and adds mass to the craft. Some spacecraft travel to far-distant regions with the help of gravity from the planets they pass by. This is a technique called gravity assist. You can model gravity assist using a simple table tennis ball.

1. Read and complete a lab safety form.
2. Set a **turntable** in motion.
3. Gently throw a **table tennis ball** so that it just skims the top of the spinning surface. You might have to practice before you're able to get the ball to glide over the surface.
4. In your Science Journal, describe or draw a picture of what you observed.

Think About This

1. Use your observations to describe how this activity is similar to gravity assist.
2. **Key Concept** How do you think gravity assist helps scientists learn about the solar system?

Missions to the Sun and the Moon

What is the future for space exploration? Scientists at NASA and other space agencies around the world have cooperatively developed goals for future space exploration. One goal is to expand human space travel within the solar system. Two steps leading to this goal are sending probes to the Sun and sending probes to the Moon.

Key Concept Check What is a goal of future space exploration?

Solar Probes

The Sun emits high-energy radiation and charged particles. Storms on the Sun can eject powerful jets of gas and charged particles into space, as shown in **Figure 15.** The Sun's high-energy radiation and charged particles can harm astronauts and damage spacecraft. To better understand these hazards, scientists study data collected by solar probes that orbit the Sun. The solar probe *Ulysses,* launched in 1990, orbited the Sun and gathered data for 19 years.

Lunar Probes

NASA and other space agencies also plan to send several probes to the Moon. The *Lunar Reconnaissance Orbiter,* launched in 2009, collects data that will help scientists select the best location for a future lunar outpost.

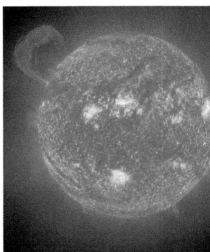

Figure 15 Storms on the Sun send charged particles far into space.

FOLDABLES

Use a sheet of copy paper to make a vertical three-tab Foldable. Draw a Venn diagram on the front tabs and use it to compare and contrast space missions to the inner and outer planets.

Missions to the Inner Planets

The inner planets are the four rocky planets closest to the Sun—Mercury, Venus, Earth, and Mars. Scientists have sent many probes to the inner planets, and more are planned. These probes help scientists learn how the inner planets formed, what geologic forces are active on them, and whether any of them could support life. Some recent and current missions to the inner planets are described in Figure 16.

Reading Check What do scientists want to learn about the inner planets?

Planetary Missions

Figure 16 Studying the solar system remains a major goal of space exploration.

◀ *Messenger* The first probe to visit Mercury—the planet closest to the Sun—since *Mariner 10* flew by the planet in 1975 is *Messenger*. After a 2004 launch and two passes of Venus, *Messenger* will fly past Mercury several times before entering Mercury's orbit in 2011. *Messenger* will study Mercury's geology and chemistry. It will send images and data back to Earth for one Earth year. On its first pass by Mercury, in 2008, *Messenger* returned over 1,000 images in many wavelengths.

Spirit and *Opportunity* Since the first flyby reached Mars in 1964, many probes have been sent to Mars. One of them produced the spectacular photo shown at the beginning of this lesson. In 2003, two robotic rovers, *Spirit* and *Opportunity*, began exploring the Martian surface for the first time. These solar-powered rovers traveled more than 20 km and relayed data for over 5 years. They have sent thousands of images to Earth. ▶

Missions to the Outer Planets and Beyond

The outer planets are the four large planets farthest from the Sun—Jupiter, Saturn, Uranus, and Neptune. Pluto was once considered an outer planet, but it is now included with other small, icy **dwarf planets** observed orbiting the Sun outside the orbit of Neptune. Missions to outer planets are long and difficult because the planets are so far from Earth. Some missions to the outer planets and beyond are described in **Figure 16** below. The next major mission to the outer planets will be an international mission to Jupiter and its four largest moons.

REVIEW VOCABULARY
dwarf planet
a round body that orbits the Sun but is not massive enough to clear away other objects in its orbit

Reading Check Why are missions to the outer planets difficult?

Visual Check Which planet has been explored by rovers?

Cassini The first orbiter sent to Saturn, *Cassini* was launched in 1997 as part of an international effort involving 19 countries. *Cassini* traveled for 7 years before entering Saturn's orbit in 2004. When it arrived, it sent a smaller probe to the surface of Saturn's largest moon, Titan, as shown at left. *Cassini* is so large—6,000 kg—that no rocket was powerful enough to send it directly to Saturn. Scientists used gravity from closer planets—Venus, Earth, and Jupiter—to help power the trip. The gravity from each planet gave the spacecraft a boost toward Saturn.

New Horizons A much smaller spacecraft, *New Horizons,* is speeding toward Pluto. *New Horizons* is also using gravity assist for its journey, with a swing past Jupiter. Launched in 2006, *New Horizons* won't reach Pluto until 2015. It will leave the solar system in 2029. Without a gravity assist from Jupiter, it would take *New Horizons* 5 years longer to reach Pluto. ▶

Figure 17 This inflatable structure could serve as housing for astronauts. It has been tested in the harsh environment of Antarctica.

ACADEMIC VOCABULARY

option
(noun) something that can be chosen

Future Space Missions

Do you think there will ever be cities or communities built outside Earth? That is looking very far ahead. No person has ever been farther than the Moon. But human space travel remains a goal of NASA and other space agencies around the world.

Studying and Visiting Mars

A visit to Mars will probably not occur for several more decades. To prepare for a visit to Mars, NASA plans to send additional probes to the planet. The probes will explore sites on Mars that might have resources that can support life. One of these probes is the *MAVEN* spacecraft. *MAVEN* will study the atmosphere of Mars and how it has evolved over time.

Astronauts will need secure housing once they establish a suitable landing area on Mars. The structure in **Figure 17** is one of those options.

Studying Jupiter

The largest planet in the solar system is going to be studied by the *Juno* spacecraft. It will take *Juno* five years to reach the gas giant. *Juno* will study Jupiter's atmosphere, gravity, magnetic fields, and atmosphere conditions.

Inquiry MiniLab   20 minutes

What conditions are required for life on Earth?
Billions of organisms live on Earth. What are the requirements for life?

1. Observe a **terrarium.** In your Science Journal, make a sketch of this environment and label every component as either living or nonliving.

2. Observe an **aquarium.** Again, make a sketch of this environment and label every component as living or nonliving.

Analyze and Conclude

1. **Compare and Contrast** Describe what the organisms in both environments need to survive.

2. **Draw Conclusions** Do all living things have the same needs? Support your answer using examples from your observations.

3. 🗝 **Key Concept** What conditions are required for life on Earth? How would knowing these requirements help scientists look for life in space?

The Search for Life

No one knows if life exists beyond Earth, but people have thought about the possibility for a long time. It even has a name. *Life that originates outside Earth is* **extraterrestrial** (ek struh tuh RES tree ul) **life.**

Conditions Needed for Life

Astrobiology *is the study of life in the universe, including life on Earth and the possibility of extraterrestrial life.* Investigating the conditions for life on Earth helps scientists predict where they might find life elsewhere in the solar system. Astrobiology also can help scientists locate environments in space where humans and other Earth life might be able to survive.

Life exists in a wide range of environments on Earth. Life-forms survive on dark ocean floors, deep in solid rocks, and in scorching water, such as the hot spring shown in **Figure 18.** No matter how extreme their environments, all known life-forms on Earth need liquid water, organic molecules, and some source of energy to survive. Scientists assume that if life exists elsewhere in space it would have the same requirements.

 Key Concept Check What is required for life on Earth?

Water in the Solar System

A lunar space probe found water in a crater on the Moon. Enough frozen water was found in a single crater to fill 1,500 Olympic swimming pools. Evidence from other space probes suggests that water vapor or ice exists on many planets and moons in the solar system. NASA plans to launch the *Mars Science Laboratory* to sample a variety of soils and rocks on Mars. This mission will investigate the possibility that life exists or once existed on the planet.

Some of the moons in the outer solar system, such as Jupiter's moon Europa, shown in **Figure 19,** might also have large amounts of liquid water beneath their surfaces.

▲ **Figure 18** Bacteria live in the boiling water of this hot spring in Yellowstone National Park.

WORD ORIGIN

astrobiology
from Greek *astron*, means "star"; Greek *bios*, means "life"; and Greek *logia*, means "study"

Figure 19 The dark patches in the inset photo might represent areas where water from an underground ocean has seeped to Europa's surface. ▼

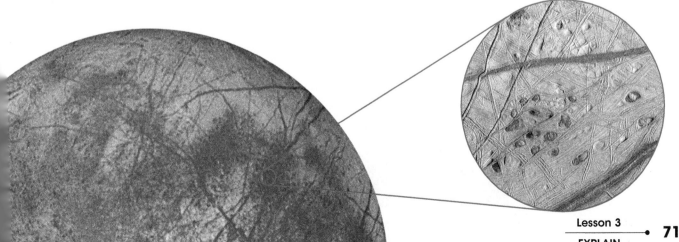

▲ Figure 20 *Kepler* is orbiting the Sun, searching a single area of sky for Earthlike planets.

Understanding Earth by Exploring Space

Space provides frontiers for the human spirit of exploration and discovery. The exploration of space also provides insight into planet Earth. Information gathered in space helps scientists understand how the Sun and other bodies in the solar system influence Earth, how Earth formed, and how Earth supports life. Looking for Earthlike planets outside the solar system helps scientists learn if Earth is unique in the universe.

Searching for Other Planets

Astronomers have detected more than 300 planets outside the solar system. Most of these planets are much bigger than Earth and probably could not support liquid water—or life. To search for Earthlike planets, NASA launched the *Kepler* telescope in 2009. The *Kepler* telescope, illustrated in Figure 20, focuses on a single area of sky containing about 100,000 stars. However, though it might detect Earthlike planets orbiting other stars, *Kepler* will not be able to detect life on any planet.

Understanding Our Home Planet

Not all of NASA's missions are to other planets, to other moons, or to look at stars and galaxies. NASA and space agencies around the world also launch and maintain Earth-observing satellites. Satellites that orbit Earth provide large-scale images of Earth's surface. These images help scientists understand Earth's climate and weather. Figure 21 is a 2005 satellite image showing changes in ocean temperature associated with Hurricane Katrina, one of the deadliest storms in U.S. history.

 Key Concept Check How can exploring space help scientists learn about Earth?

Figure 21 Earth-orbiting satellites collect data in many wavelengths. This satellite image of Hurricane Katrina was made with a microwave sensor. ▶

✓ **Visual Check** Which part of the United States did Hurricane Katrina affect?

Lesson 3 Review

Visual Summary

The *New Horizons* spacecraft will reach Pluto in 2015.

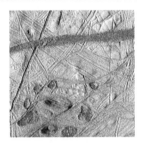

Scientists think liquid water might exist on or below the surfaces of Mars and some moons.

Earth-orbiting satellites help scientists understand weather and climate patterns on Earth.

FOLDABLES

Use your lesson Foldable to review the lesson. Save your Foldable for the project at the end of the chapter.

What do you think NOW?

You first read the statements below at the beginning of the chapter.

5. Humans have landed on Mars.

6. Scientists have detected water on other bodies in the solar system.

Did you change your mind about whether you agree or disagree with the statements? Rewrite any false statements to make them true.

Use Vocabulary

1. **Use the term** *extraterrestrial life* in a sentence.

2. The study of life in the universe is _____.

Understand Key Concepts

3. Which gave *Cassini* a boost toward Saturn?
   A. buoyancy   C. magnetism
   B. gravity   D. wind

4. **Explain** why bodies that have liquid water are the best candidates for supporting life.

5. **Assess** the benefits of an inflatable structure over a concrete structure on the Moon.

6. **Identify** some phenomena on Earth best viewed by artificial satellites.

Interpret Graphics

7. **Assess** The figure above represents a possible design for a new solar probe that would orbit close to the Sun. What purpose might the part labeled *A* serve?

8. **Organize Information** Copy and fill in the graphic organizer below to list requirements for life on Earth.

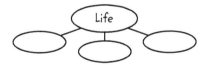

Critical Thinking

9. **Predict** some of the challenges people might face living in a lunar outpost.

10. **Debate** whether scientists should look first for life on Mars or on Europa.

Inquiry Lab

2 class periods

Materials

newspaper

creative building materials

masking tape

cup varieties

office supplies

craft supplies

scissors

Safety

Design and Construct a Moon Habitat

No one has visited the Moon since 1972. NASA plans to send astronauts back to the Moon as early as 2020. You might be one of the lucky ones who will be sent to find a suitable location for a lunar outpost. To get a head start, your task is to design and build a model of a moon habitat where people can live and work for months at a time. You can use any materials provided or other materials approved by your teacher. Before you begin, think about some of the things people will need in order to survive on the Moon.

Question

Think about what humans need on a daily basis. How can you design a moon habitat that would meet people's needs in a place very unlike Earth?

Procedure

1. Read and complete a lab safety form.

2. Think about construction. Consider the function each material might represent in a moon habitat. The materials will have to be transported from Earth to the Moon before any construction can begin.

3. Draw plans for your moon habitat. Be sure to include an airlock, a small room that separates an outer door from an inner door. Label the materials you will use and what each represents.

4. Copy the data table below into your Science Journal. Complete the table by listing each material you plan to use, its purpose or function, and why you chose it.

Materials for a Moon Habitat		
Material	Function	Why I Chose the Material

714 Chapter 19 EXTEND

Lab Tips

☑ Before you begin, make a list of conditions on the Moon that are much different from those on Earth.

☑ If you can think of any materials not listed that you would like to use, ask your teacher's permission to use them.

Analyze and Conclude

8. **Explain** in detail why you chose the materials and the design that you did.
9. **Evaluate** Which materials or designs did not work as expected? Explain.
10. **Compare and Contrast** What differences between the lunar environment and Earth's environment did you consider in your design?
11. **The Big Idea** What requirements must be met for humans to live, work, and be healthy on the Moon?

Communicate Your Results

Imagine that your design is part of a NASA competition to find the best lunar habitat. Write and give a 2–3 minute presentation convincing NASA to use your model for its moon habitat.

5. Build your moon habitat. When you are finished, check to see that your habitat satisfies the conditions in your original question. If not, revise your habitat or make a note in your Science Journal about how you would improve it.

 Extension

Compare your moon habitat to the habitats of at least three other groups. Discuss how you might combine your ideas to build a bigger and better moon habitat.

6. In addition to meeting people's needs in space, the habitat should be easy to construct in the harsh environments of space. Remember that the materials should be easy to transport from Earth to the Moon.

7. Some things might not go as planned as you construct your model, or you might get new ideas as you proceed with building. As you go along, you can adapt your structure to improve the final product. Record any changes you make to your design or materials in your Science Journal.

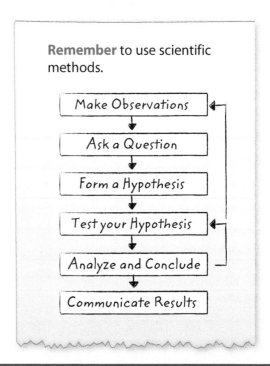

Remember to use scientific methods.

- Make Observations
- Ask a Question
- Form a Hypothesis
- Test your Hypothesis
- Analyze and Conclude
- Communicate Results

Lesson 3
EXTEND

Chapter 19 Study Guide

THE BIG IDEA

Humans observe the universe with Earth-based and space-based telescopes. They explore the solar system with crewed and uncrewed space probes.

Key Concepts Summary

Lesson 1: Observing the Universe

- Scientists use different parts of the **electromagnetic spectrum** to study stars and other objects in space and to learn what the universe was like many millions of years ago.
- Telescopes in space can collect radiant energy that Earth's atmosphere would absorb or refract.

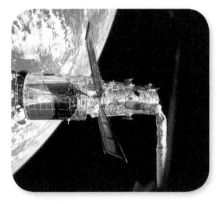

Lesson 2: Early History of Space Exploration

- **Rockets** are used to overcome the force of Earth's gravity when sending **satellites, space probes,** and other spacecraft into space.
- Uncrewed missions can make trips that are too long or too dangerous for humans.
- Materials and technologies from the space program have been applied to everyday life.

Lesson 3: Recent and Future Space Missions

- A goal of the space program is to expand human space travel within the solar system and develop lunar and Martian outposts.
- All known life-forms need liquid water, energy, and organic molecules.
- Information gathered in space helps scientists understand how the Sun influences Earth, how Earth formed, whether life exists outside of Earth, and how weather and climate affect Earth.

Vocabulary

electromagnetic spectrum p. 690
refracting telescope p. 692
reflecting telescope p. 692
radio telescope p. 693

rocket p. 699
satellite p. 700
space probe p. 701
lunar p. 701
Project Apollo p. 702
space shuttle p. 702

extraterrestrial life p. 711
astrobiology p. 711

Study Guide

Review
- Personal Tutor
- Vocabulary eGames
- Vocabulary eFlashcards

FOLDABLES Chapter Project

Assemble your lesson Foldables as shown to make a Chapter Project. Use the project to review what you have learned in this chapter.

Use Vocabulary

1. All radiation is classified by wavelength in the _____.
2. Two types of telescopes that collect visible light are _____ and _____.
3. The space mission that sent the first humans to the Moon was _____.
4. An example of a human space transportation system is a(n) _____.
5. An uncrewed spacecraft is a(n) _____.
6. The discipline that investigates life in the universe is _____.
7. The best place to find _____ is on solar system bodies containing water.

Link Vocabulary and Key Concepts

 Concepts in Motion   Interactive Concept Map

Copy this concept map, and then use vocabulary terms from the previous page to complete the concept map.

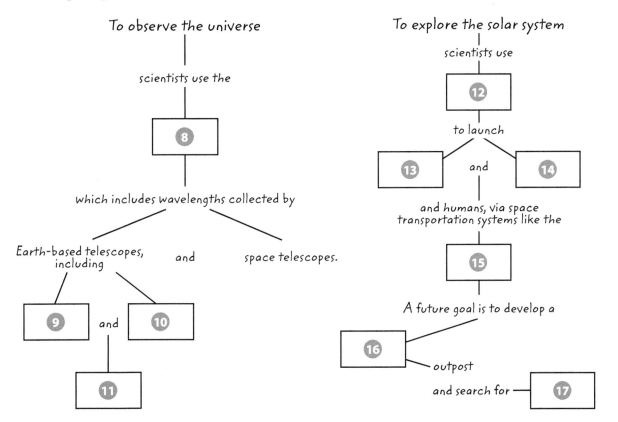

Chapter 19 Review

Understand Key Concepts

1. Which type of telescope is shown in the figure below?

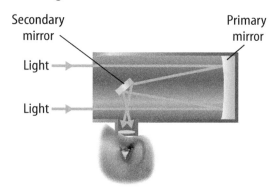

   A. infrared telescope
   B. radio telescope
   C. reflecting telescope
   D. refracting telescope

2. In which wavelength would you expect the hottest stars to emit most of their energy?
   A. gamma rays
   B. microwaves
   C. radio waves
   D. visible light

3. Which best describes *Hubble*?
   A. infrared telescope
   B. radio telescope
   C. refracting telescope
   D. space telescope

4. What is special about the *Kepler* mission?
   A. *Kepler* can detect objects at all wavelengths.
   B. *Kepler* has found the most distant objects in the universe.
   C. *Kepler* is dedicated to finding Earthlike planets.
   D. *Kepler* is the first telescope to orbit the Sun.

5. Where is the *International Space Station?*
   A. on Mars
   B. on the Moon
   C. orbiting Earth
   D. orbiting the Sun

6. Which mission sent people to the Moon?
   A. Apollo
   B. Explorer
   C. Galileo
   D. Pioneer

7. Which are most likely to have liquid water?
   A. Mars and Europa
   B. Mars and Venus
   C. the Moon and Europa
   D. the Moon and Mars

8. The images below were taken by a rover as it moved along a rocky body in the inner solar system in 2004. Which body is it?

   A. Europa
   B. Mars
   C. Titan
   D. Venus

9. Which is NOT a satellite?
   A. a flyby
   B. a moon
   C. an orbiter
   D. space telescope

Chapter Review

Critical Thinking

10 Contrast waves in the electromagnetic spectrum with water waves in the ocean.

11 Differentiate If you wanted to study new stars forming inside a huge dust cloud, which wavelength might you use? Explain.

12 Deduce Why do Earth-based optical telescopes work best at night, while radio telescopes work all day and all night long?

13 Analyze Why it is more challenging to send space probes to the outer solar system than to the inner solar system?

14 Create a list of requirements that must be satisfied before humans can live on the Moon.

15 Choose a body in the solar system that you think would be a good place to look for life. Explain.

16 Interpret Graphics Copy the diagram of electromagnetic waves below, and label the relative positions of ultraviolet waves, X-rays, visible light, infrared waves, microwaves, gamma rays, and radio waves.

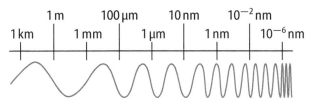

Writing in Science

17 Write a paragraph comparing colonizing North America and colonizing the Moon. Include a main idea, supporting details, and a concluding sentence.

REVIEW THE BIG IDEA

18 In what different ways do humans observe and explore space?

19 The photo below shows the *Hubble Space Telescope* orbiting Earth. What are advantages of space-based telescopes? What are disadvantages?

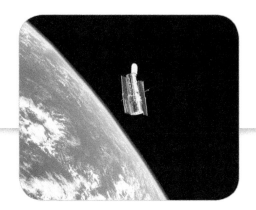

Math Skills

Use Scientific Notation

20 The distance from Saturn to the Sun averages 1,430,000,000 km. Express this distance in scientific notation.

21 The nearest star outside our solar system is Proxima Centauri, which is about 39,900,000,000,000 km from Earth. What is this distance in scientific notation?

22 The *Hubble Space Telescope* has taken pictures of an object that is 1,400,000,000,000,000,000,000 km away from Earth. Express this number in scientific notation.

Standardized Test Practice

Record your answers on the answer sheet provided by your teacher or on a sheet of paper.

Multiple Choice

1. Which is NOT a good place to build a radio telescope?
   - A a location near a radio station
   - B a location that is remote
   - C a location with a large cleared area
   - D a location with dry air

2. Which has the power to overcome the force of Earth's gravity to be launched into space?
   - A a probe
   - B a rocket
   - C a satellite
   - D a telescope

Use the figure below to answer question 3.

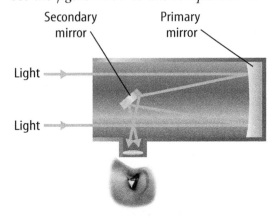

3. Which could increase the light-gathering power of the telescope in the figure?
   - A adaptive optics
   - B a larger eyepiece
   - C multiple small mirrors
   - D thicker lenses

4. Which lists the minimum resources needed for life-forms to survive on Earth?
   - A liquid water, an energy source, and sunshine
   - B liquid water, sunshine, and organic molecules
   - C organic molecules, an energy source, and liquid water
   - D organic molecules, an energy source, and sunshine

Use the table below to answer questions 5 and 6.

Planet	Average Distance from Sun (in millions of kilometers)
Earth	150
Mars	228
Saturn	1,434

5. It takes about 8.3 min for light to travel from the Sun to Earth. It takes about 40 min for light to travel from the Sun to Jupiter. How long would you expect it to take light to travel from the Sun to Saturn?
   - A 8.5 min
   - B 1.3 h
   - C 13.5 h
   - D 26.3 h

6. Which shows the distance between Saturn and the Sun expressed in scientific notation?
   - A 1.434×10^6 km
   - B 1.434×10^8 km
   - C 1.434×10^9 km
   - D 14.34×10^7 km

Standardized Test Practice

7 What is the advantage of using gravity assist for a mission to Saturn?
   A The spacecraft can be made of a nonmagnetic material.
   B The spacecraft can travel at the speed of light.
   C The spacecraft needs less fuel.
   D The spacecraft needs more weight.

8 Which was the first satellite to orbit Earth?
   A *Apollo 1*
   B *Explorer 1*
   C *Mariner 1*
   D *Sputnik 1*

Use the figure below to answer question 9.

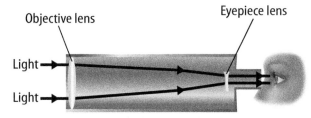

9 Which is true of the telescope above?
   A The eyepiece and the objective lens are concave lenses.
   B Light is bent as it goes through the objective lens.
   C Light is reflected from the eyepiece lens to the objective lens.
   D The eyepiece lens can be made of many smaller lenses.

Constructed Response

Use the figure below to answer questions 10 and 11.

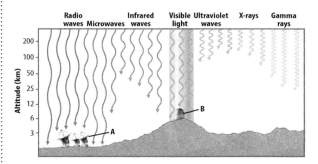

10 Identify the types of telescopes labeled *A* and *B* in the figure. Briefly explain what radiant energy each gathers and how each telescope works.

11 Use the information in the figure to explain why X-ray images can be obtained only using telescopes located above Earth's atmosphere.

12 How does studying radiant energy help scientists learn about the universe?

13 How might the properties of materials developed for use in space be useful on Earth? Give examples.

14 How does information gathered in space help scientists learn about Earth?

15 How does the *Kepler telescope* differ from other telescopes in space?

NEED EXTRA HELP?

If You Missed Question...	1	2	3	4	5	6	7	8	9	10	11	12	13	14	15
Go to Lesson...	1	2	1	3	1	1	3	2	1	1	1	1	2	3	3

Chapter 20

The Sun-Earth-Moon System

THE BIG IDEA — What natural phenomena do the motions of Earth and the Moon produce?

Inquiry Sun Bites?

Look at this time-lapse photograph. The "bites" out of the Sun occurred during a solar eclipse. The Sun's appearance changed in a regular, predictable way as the Moon's shadow passed over a part of Earth.

- How does the Moon's movement change the Sun's appearance?
- What predictable changes does Earth's movement cause?
- What other natural phenomena do the motions of Earth and the Moon cause?

Get Ready to Read

What do you think?
Before you read, decide if you agree or disagree with each of these statements. As you read this chapter, see if you change your mind about any of the statements.

1. Earth's movement around the Sun causes sunrises and sunsets.
2. Earth has seasons because its distance from the Sun changes throughout the year.
3. The Moon was once a planet that orbited the Sun between Earth and Mars.
4. Earth's shadow causes the changing appearance of the Moon.
5. A solar eclipse happens when Earth moves between the Moon and the Sun.
6. The gravitational pull of the Moon and the Sun on Earth's oceans causes tides.

ConnectED — Your one-stop online resource

connectED.mcgraw-hill.com

- Video
- WebQuest
- Audio
- Assessment
- Review
- Concepts in Motion
- Inquiry
- Multilingual eGlossary

Lesson 1

Earth's Motion

Reading Guide

Key Concepts
ESSENTIAL QUESTIONS

- How does Earth move?
- Why is Earth warmer at the equator and colder at the poles?
- Why do the seasons change as Earth moves around the Sun?

Vocabulary

orbit p. 726
revolution p. 726
rotation p. 727
rotation axis p. 727
solstice p. 731
equinox p. 731

 Multilingual eGlossary

Inquiry Floating in Space?

From the *International Space Station*, Earth might look like it is just floating, but it is actually traveling around the Sun at more than 100,000 km/h. What phenomena does Earth's motion cause?

724 • Chapter 20
ENGAGE

Inquiry Launch Lab
15 minutes

Does Earth's shape affect temperatures on Earth's surface?
Temperatures near Earth's poles are colder than temperatures near the equator. What causes these temperature differences?

1. Read and complete a lab safety form.
2. Inflate a **spherical balloon** and tie the balloon closed.
3. Using a **marker**, draw a line around the balloon to represent Earth's equator.
4. Using a **ruler**, place a lit **flashlight** about 8 cm from the balloon so the flashlight beam strikes the equator straight on.
5. Using the marker, trace around the light projected onto the balloon.
6. Have someone raise the flashlight vertically 5–8 cm without changing the direction that the flashlight is pointing. Do not change the position of the balloon. Trace around the light projected onto the balloon again.

Think About This
1. Compare and contrast the shapes you drew on the balloon.
2. At which location on the balloon is the light more spread out? Explain your answer.
3. **Key Concept** Use your model to explain why Earth is warmer near the equator and colder near the poles.

Earth and the Sun

If you look outside at the ground, trees, and buildings, it does not seem like Earth is moving. Yet Earth is always in motion, spinning in space and traveling around the Sun. As Earth spins, day changes to night and back to day again. The seasons change as Earth travels around the Sun. Summer changes to winter because Earth's motion changes how energy from the Sun spreads out over Earth's surface.

The Sun

The nearest star to Earth is the Sun, which is shown in **Figure 1**. The Sun is approximately 150 million km from Earth. Compared to Earth, the Sun is enormous. The Sun's diameter is more than 100 times greater than Earth's diameter. The Sun's mass is more than 300,000 times greater than Earth's mass.

Deep inside the Sun, nuclei of atoms combine, releasing huge amounts of energy. This process is called nuclear fusion. The Sun releases so much energy from nuclear fusion that the temperature at its core is more than 15,000,000°C. Even at the Sun's surface, the temperature is about 5,500°C. A small part of the Sun's energy reaches Earth as light and thermal energy.

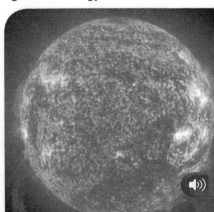

Figure 1 The Sun is a giant ball of hot gases that emits light and energy.

Inquiry MiniLab — 10 minutes

What keeps Earth in orbit?
Why does Earth move around the Sun and not fly off into space?

1. Read and complete a lab safety form.
2. Tie a piece of **strong thread** securely to a **plastic, slotted golf ball.**
3. Swing the ball in a horizontal circle above your head.

Analyze and Conclude

1. **Predict** what would happen if you let go of the thread.
2. **Key Concept** Which part of the experiment represents the force of gravity between Earth and the Sun?

Earth's Orbit

As shown in **Figure 2,** Earth moves around the Sun in a nearly circular path. *The path an object follows as it moves around another object is an* **orbit.** *The motion of one object around another object is called* **revolution.** Earth makes one complete revolution around the Sun every 365.24 days.

The Sun's Gravitational Pull

Why does Earth orbit the Sun? The answer is that the Sun's gravity pulls on Earth. The pull of gravity between two objects depends on the masses of the objects and the distance between them. The more mass either object has, or the closer together they are, the stronger the gravitational pull.

The Sun's effect on Earth's motion is illustrated in **Figure 2.** Earth's motion around the Sun is like the motion of an object twirled on a string. The string pulls on the object and makes it move in a circle. If the string breaks, the object flies off in a straight line. In the same way, the pull of the Sun's gravity keeps Earth revolving around the Sun in a nearly circular orbit. If the gravity between Earth and the Sun were to somehow stop, Earth would fly off into space in a straight line.

Key Concept Check What produces Earth's revolution around the Sun?

Figure 2 Earth moves in a nearly circular orbit. The pull of the Sun's gravity on Earth causes Earth to revolve around the Sun.

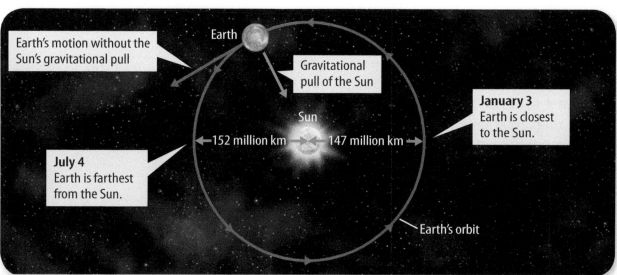

Earth's Rotation Axis

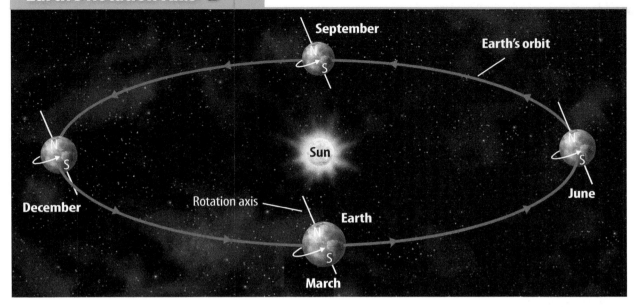

Figure 3 This diagram shows Earth's orbit, which is nearly circular, from an angle. Earth spins on its rotation axis as it revolves around the Sun. Earth's rotation axis always points in the same direction.

Visual Check Between which months is the north end of Earth's rotation axis away from the Sun?

Earth's Rotation

As Earth revolves around the Sun, it spins. *A spinning motion is called* **rotation**. Some spinning objects rotate on a rod or axle. Earth rotates on an imaginary line through its center. *The line on which an object rotates is the* **rotation axis.**

Suppose you could look down on Earth's North Pole and watch Earth rotate. You would see that Earth rotates on its rotation axis in a counterclockwise direction, from west to east. One complete rotation of Earth takes about 24 hours. This rotation helps produce Earth's cycle of day and night. It is daytime on the half of Earth facing toward the Sun and nighttime on the half of Earth facing away from the Sun.

The Sun's Apparent Motion Each day the Sun appears to move from east to west across the sky. It seems as if the Sun is moving around Earth. However, it is Earth's rotation that causes the Sun's apparent motion.

Earth rotates from west to east. As a result, the Sun appears to move from east to west across the sky. The stars and the Moon also seem to move from east to west across the sky due to Earth's west to east rotation.

To better understand this, imagine riding on a merry-go-round. As you and the ride move, people on the ground appear to be moving in the opposite direction. In the same way, as Earth rotates from west to east, the Sun appears to move from east to west.

Reading Check What causes the Sun's apparent motion across the sky?

The Tilt of Earth's Rotation Axis As shown in **Figure 3,** Earth's rotation axis is tilted. The tilt of Earth's rotation axis is always in the same direction by the same amount. This means that during half of Earth's orbit, the north end of the rotation axis is toward the Sun. During the other half of Earth's orbit, the north end of the rotation axis is away from the Sun.

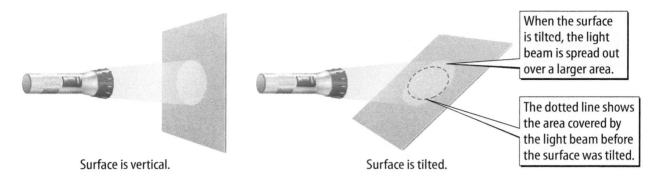

Surface is vertical.   Surface is tilted.

When the surface is tilted, the light beam is spread out over a larger area.

The dotted line shows the area covered by the light beam before the surface was tilted.

Figure 4 The light energy on a surface becomes more spread out as the surface becomes more tilted relative to the light beam.

 Visual Check Is the light energy more spread out on the vertical or tilted surface?

Temperature and Latitude

As Earth orbits the Sun, only one half of Earth faces the Sun at a time. A beam of sunlight carries energy. The more sunlight that reaches a part of Earth's surface, the warmer that part becomes. Because Earth's surface is curved, different parts of Earth's surface receive different amounts of the Sun's energy.

Energy Received by a Tilted Surface

Suppose you shine a beam of light on a flat card, as shown in **Figure 4.** As you tilt the card relative to the direction of the light beam, light becomes more spread out on the card's surface. As a result, the energy that the light beam carries also spreads out more over the card's surface. An area on the surface within the light beam receives less energy when the surface is more tilted relative to the light beam.

The Tilt of Earth's Curved Surface

Instead of being flat like a card, Earth's surface is curved. Relative to the direction of a beam of sunlight, Earth's surface becomes more tilted as you move away from the equator. As shown in **Figure 5,** the energy in a beam of sunlight tends to become more spread out the farther you travel from the equator. This means that regions near the poles receive less energy than regions near the equator. This makes Earth colder at the poles and warmer at the equator.

Key Concept Check Why is Earth warmer at the equator and colder at the poles?

ACADEMIC VOCABULARY

equator
(noun) the imaginary line that divides Earth into its northern and southern hemispheres

Figure 5 Energy from the Sun becomes more spread out as you move away from the equator.

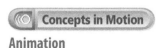
Animation

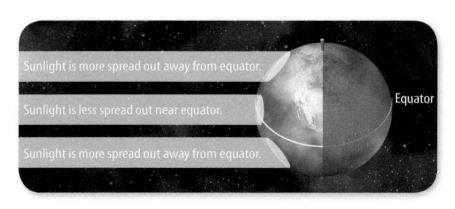

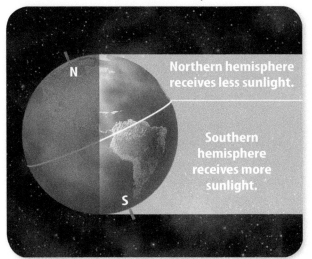
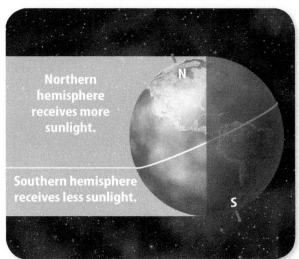

Figure 6 The northern hemisphere receives more sunlight in June, and the southern hemisphere receives more sunlight in December.

Seasons

You might think that summer happens when Earth is closest to the Sun, and winter happens when Earth is farthest from the Sun. However, seasonal changes do not depend on Earth's distance from the Sun. In fact, Earth is closest to the Sun in January! Instead, it is the tilt of Earth's rotation axis, combined with Earth's motion around the Sun, that causes the seasons to change.

Spring and Summer in the Northern Hemisphere

During one half of Earth's orbit, the north end of the rotation axis is toward the Sun. Then, the northern hemisphere receives more energy from the Sun than the southern hemisphere, as shown in Figure 6. Temperatures increase in the northern hemisphere and decrease in the southern hemisphere. Daylight hours last longer in the northern hemisphere, and nights last longer in the southern hemisphere. This is when spring and summer happen in the northern hemisphere, and fall and winter happen in the southern hemisphere.

Fall and Winter in the Northern Hemisphere

During the other half of Earth's orbit, the north end of the rotation axis is away from the Sun. Then, the northern hemisphere receives less solar energy than the southern hemisphere, as shown in Figure 6. Temperatures decrease in the northern hemisphere and increase in the southern hemisphere. This is when fall and winter happen in the northern hemisphere, and spring and summer happen in the southern hemisphere.

Key Concept Check How does the tilt of Earth's rotation axis affect Earth's weather?

Math Skills

Convert Units
When Earth is 147,000,000 km from the Sun, how far is Earth from the Sun in miles? To calculate the distance in miles, multiply the distance in km by the conversion factor

$$147{,}000{,}000 \text{ km} \times \frac{0.62 \text{ miles}}{1 \text{ km}}$$
$$= 91{,}100{,}000 \text{ miles}$$

Practice
When Earth is 152,000,000 km from the Sun, how far is Earth from the Sun in miles?

- Math Practice
- Personal Tutor

Earth's Seasonal Cycle

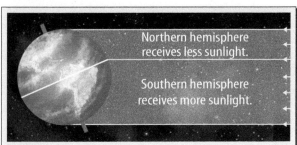

December Solstice
The December solstice is on December 21 or 22. On this day
- the north end of Earth's rotation axis is away from the Sun;
- days in the northern hemisphere are shortest and nights are longest; winter begins;
- days in the southern hemisphere are longest and nights are shortest; summer begins.

September Equinox
The September equinox is on September 22 or 23. On this day
- the north end of Earth's rotation axis leans along Earth's orbit;
- there are about 12 hours of daylight and 12 hours of darkness everywhere on Earth;
- autumn begins in the northern hemisphere;
- spring begins in the southern hemisphere.

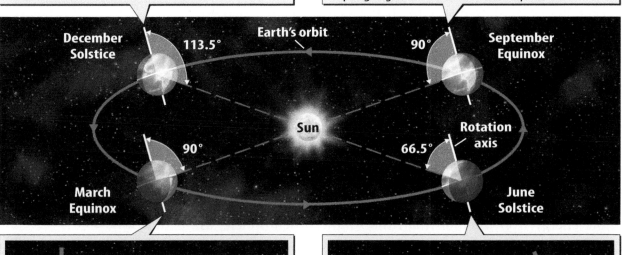

March Equinox
The March equinox is on March 20 or 21. On this day
- the north end of Earth's rotation axis leans along Earth's orbit;
- there are about 12 hours of daylight and 12 hours of darkness everywhere on Earth;
- spring begins in the northern hemisphere;
- autumn begins in the southern hemisphere.

June Solstice
The June solstice is on June 20 or 21. On this day
- the north end of Earth's rotation axis is toward the Sun;
- days in the northern hemisphere are longest and nights are shortest; summer begins;
- days in the southern hemisphere are shortest and nights are longest; winter begins.

Figure 7 The seasons change as Earth moves around the Sun. Earth's motion around the Sun causes Earth's tilted rotation axis to be leaning toward the Sun and away from the Sun.

Solstices, Equinoxes, and the Seasonal Cycle

Figure 7 shows that as Earth travels around the Sun, its rotation axis always points in the same direction in space. However, the amount that Earth's rotation axis is toward or away from the Sun changes. This causes the seasons to change in a yearly cycle.

There are four days each year when the direction of Earth's rotation axis is special relative to the Sun. A **solstice** *is a day when Earth's rotation axis is the most toward or away from the Sun.* An **equinox** *is a day when Earth's rotation axis is leaning along Earth's orbit, neither toward nor away from the Sun.*

March Equinox to June Solstice When the north end of the rotation axis gradually points more and more toward the Sun, the northern hemisphere gradually receives more solar energy. This is spring in the northern hemisphere.

June Solstice to September Equinox The north end of the rotation axis continues to point toward the Sun but does so less and less. The northern hemisphere starts to receive less solar energy. This is summer in the northern hemisphere.

September Equinox to December Solstice The north end of the rotation axis now points more and more away from the Sun. The northern hemisphere receives less and less solar energy. This is fall in the northern hemisphere.

December Solstice to March Equinox The north end of the rotation axis continues to point away from the Sun but does so less and less. The northern hemisphere starts to receive more solar energy. This is winter in the northern hemisphere.

Changes in the Sun's Apparent Path Across the Sky

Figure 8 shows how the Sun's apparent path through the sky changes from season to season in the northern hemisphere. The Sun's apparent path through the sky in the northern hemisphere is lowest on the December solstice and highest on the June solstice.

FOLDABLES
Make a bound book with four full pages. Label the pages with the names of the solstices and equinoxes. Use each page to organize information about each season.

WORD ORIGIN
equinox
from Latin *equinoxium*, means "equality of night and day"

Figure 8 As the seasons change, the path of the Sun across the sky changes. In the northern hemisphere, the Sun's path is lowest on the December solstice and highest on the June solstice.

Visual Check When is the Sun highest in the sky in the northern hemisphere?

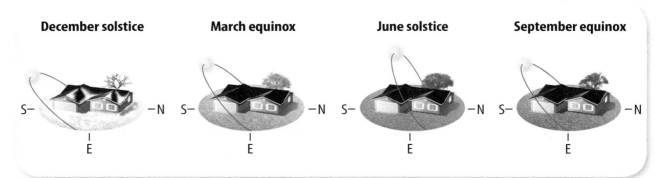

Lesson 1 Review

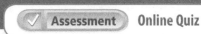 Assessment   Online Quiz

Visual Summary

 The gravitational pull of the Sun causes Earth to revolve around the Sun in a near-circular orbit.

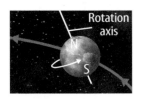 Earth's rotation axis is tilted and always points in the same direction in space.

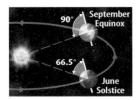 Equinoxes and solstices are days when the direction of Earth's rotation axis relative to the Sun is special.

FOLDABLES

Use your lesson Foldable to review the lesson. Save your Foldable for the project at the end of the chapter.

What do you think NOW?

You first read the statements below at the beginning of the chapter.

1. Earth's movement around the Sun causes sunrises and sunsets.
2. Earth has seasons because its distance from the Sun changes throughout the year.

Did you change your mind about whether you agree or disagree with the statements? Rewrite any false statements to make them true.

Use Vocabulary

1. **Distinguish** between Earth's rotation and Earth's revolution.

2. The path Earth follows around the Sun is Earth's _____.

3. When a(n) _____ occurs, the northern hemisphere and the southern hemisphere receive the same amount of sunlight.

Understand Key Concepts

4. What is caused by the tilt of Earth's rotational axis?
   A. Earth's orbit   C. Earth's revolution
   B. Earth's seasons   D. Earth's rotation

5. **Contrast** the amount of sunlight received by an area near the equator and a same-sized area near the South Pole.

6. **Contrast** the Sun's gravitational pull on Earth when Earth is closest to the Sun and when Earth is farthest from the Sun.

Interpret Graphics

7. **Summarize** Copy and fill in the table below for the seasons in the northern hemisphere.

Season	Starts on Solstice or Equinox?	How Rotation Axis Leans
Summer		
Fall		
Winter		
Spring		

Critical Thinking

8. **Defend** The December solstice is often called the winter solstice. Do you think this is an appropriate label? Defend your answer.

Math Skills

Review — Math Practice

9. The Sun's diameter is about 1,390,000 km. What is the Sun's diameter in miles?

732 • Chapter 20
EVALUATE

Inquiry Skill Practice: Draw Conclusions

25 minutes

Materials

large foam ball

wooden skewer

foam cup

masking tape

flashlight

marker

Safety

How does Earth's tilted rotation axis affect the seasons?

The seasons change as Earth revolves around the Sun. How does Earth's tilted rotation axis change how sunlight spreads out over different parts of Earth's surface?

Learn It

Using a flashlight as the Sun and a foam ball as Earth, you can model how solar energy spreads out over Earth's surface at different times during the year. This will help you **draw conclusions** about Earth's seasons.

Try It

1. Read and complete a lab safety form.

2. Insert a wooden skewer through the center of a foam ball. Draw a line on the ball to represent Earth's equator. Insert one end of the skewer into an upside-down foam cup so the skewer tilts.

3. Prop a flashlight on a stack of books about 0.5 m from the ball. Turn on the flashlight and position the ball so the skewer points toward the flashlight, representing the June solstice.

4. In your Science Journal, draw how the ball's surface is tilted relative to the light beam.

5. Under your diagram, state whether the upper (northern) or lower (southern) hemisphere receives more light energy.

6. With the skewer always pointing in the same direction, move the ball around the flashlight. Turn the flashlight to keep the light on the ball. At the three positions corresponding to the equinoxes and other solstice, make drawings like those in step 4 and statements like those in step 5.

Apply It

7. How did the tilt of the surfaces change relative to the light beam as the ball circled the flashlight?

8. How did the amount of light energy on each hemisphere change as the ball moved around the flashlight?

9. **Key Concept** Draw conclusions about how Earth's tilt affects the seasons.

Lesson 1
EXTEND

Lesson 2

Reading Guide

Key Concepts

ESSENTIAL QUESTIONS

- How does the Moon move around Earth?
- Why does the Moon's appearance change?

Vocabulary

maria p. 736
phase p. 738
waxing phase p. 738
waning phase p. 738

Multilingual eGlossary

Earth's Moon

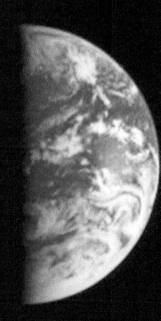

Inquiry Two Planets?

The smaller body is Earth's Moon, not a planet. Just as Earth moves around the Sun, the Moon moves around Earth. The Moon's motion around Earth causes what kinds of changes to occur?

Launch Lab

15 minutes

Why does the Moon appear to change shape?

The Sun is always shining on Earth and the Moon. However, the Moon's shape seems to change from night to night and day to day. What could cause the Moon's appearance to change?

1. Read and complete a lab safety form.
2. Place a **ball** on a level surface.
3. Position a **flashlight** so that the light beam shines fully on one side of the ball. Stand behind the flashlight.
4. Make a drawing of the ball's appearance in your Science Journal.
5. Stand behind the ball, facing the flashlight, and repeat step 4.
6. Stand to the left of the ball and repeat step 4.

Think About This

1. What caused the ball's appearance to change?
2. **Key Concept** What do you think produces the Moon's changing appearance in the sky?

Seeing the Moon

Imagine what people thousands of years ago thought when they looked up at the Moon. They might have wondered why the Moon shines and why it seems to change shape. They probably would have been surprised to learn that the Moon does not emit light at all. Unlike the Sun, the Moon is a solid object that does not emit its own light. You only see the Moon because light from the Sun reflects off the Moon and into your eyes. Some facts about the Moon, such as its mass, size, and distance from Earth, are shown in Table 1.

FOLDABLES

Use two sheets of paper to make a bound book. Use it to organize information about the lunar cycle. Each page of your book should represent one week of the lunar cycle.

Table 1 Moon Data				
Mass	Diameter	Average distance from Earth	Time for one rotation	Time for one revolution
1.2% of Earth's mass	27% of Earth's diameter	384,000 km	27.3 days	27.3 days

Lesson 2 **735**
EXPLORE

Figure 9 The Moon probably formed when a large object collided with Earth 4.5 billion years ago. Material ejected from the collision eventually clumped together and became the Moon.

Concepts in Motion Animation

An object the size of Mars crashes into the semi-molten Earth about 4.5 billion years ago.

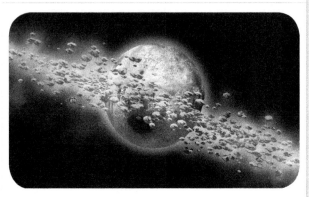

The impact ejects vaporized rock into space. As the rock cools, it forms a ring of particles around Earth.

The particles gradually clump together and form the Moon.

WORD ORIGIN
maria
from Latin *mare*, means "sea"

The Moon's Formation

The most widely accepted idea for the Moon's formation is the giant impact hypothesis, shown in **Figure 9.** According to this hypothesis, shortly after Earth formed about 4.6 billion years ago, an object about the size of the planet Mars collided with Earth. The impact ejected vaporized rock that formed a ring around Earth. Eventually, the material in the ring cooled, clumped together, and formed the Moon.

The Moon's Surface

The surface of the Moon was shaped early in its history. Examples of common features on the Moon's surface are shown in **Figure 10.**

Craters The Moon's craters were formed when objects from space crashed into the Moon. Light-colored streaks called rays extend outward from some craters.

Most of the impacts that formed the Moon's craters occurred more than 3.5 billion years ago, long before dinosaurs lived on Earth. Earth was also heavily bombarded by objects from space during this time. However, on Earth, wind, liquid water, and plate tectonics erased the craters. The Moon has no atmosphere, liquid water, or plate tectonics, so craters formed billions of years ago on the Moon have hardly changed.

Maria *The large, dark, flat areas on the Moon are called* **maria** (MAR ee uh). The maria formed after most impacts on the Moon's surface had stopped. Maria formed when lava flowed up through the Moon's crust and solidified. The lava covered many of the Moon's craters and other features. When this lava solidified, it was dark and flat.

Reading Check How were maria produced?

Highlands The light-colored highlands are too high for the lava that formed the maria to reach. The highlands are older than the maria and are covered with craters.

The Moon's Surface Features

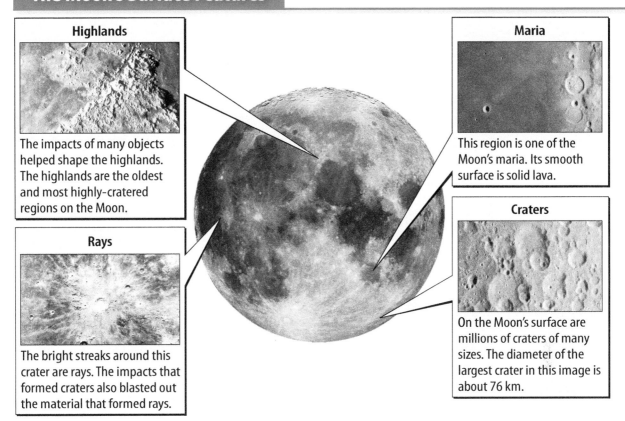

Highlands
The impacts of many objects helped shape the highlands. The highlands are the oldest and most highly-cratered regions on the Moon.

Rays
The bright streaks around this crater are rays. The impacts that formed craters also blasted out the material that formed rays.

Maria
This region is one of the Moon's maria. Its smooth surface is solid lava.

Craters
On the Moon's surface are millions of craters of many sizes. The diameter of the largest crater in this image is about 76 km.

▲ **Figure 10** The Moon's surface features include craters, rays, maria, and highlands.

The Moon's Motion

While Earth is revolving around the Sun, the Moon is revolving around Earth. The gravitational pull of Earth on the Moon causes the Moon to move in an orbit around Earth. The Moon makes one revolution around Earth every 27.3 days.

Key Concept Check What produces the Moon's revolution around Earth?

The Moon also rotates as it revolves around Earth. One complete rotation of the Moon also takes 27.3 days. This means the Moon makes one rotation in the same amount of time that it makes one revolution around Earth. Figure 11 shows that, because the Moon makes one rotation for each revolution of Earth, the same side of the Moon always faces Earth. This side of the Moon is called the near side. The side of the Moon that cannot be seen from Earth is called the far side of the Moon.

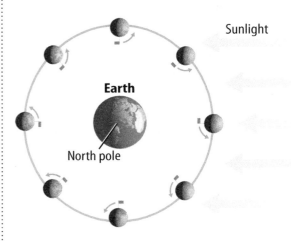

▲ **Figure 11** The Moon rotates once on its axis and revolves around Earth in the same amount of time. As a result, the same side of the Moon always faces Earth.

Inquiry MiniLab

10 minutes

How can the Moon be rotating if the same side of the Moon is always facing Earth?

The Moon revolves around Earth. Does the Moon also rotate as it revolves around Earth?

1. Choose a partner. One person represents the Moon. The other represents Earth.
2. While Earth is still, the Moon moves slowly around Earth, always facing the same wall.
3. Next, the Moon moves around Earth always facing Earth.

Analyze and Conclude

1. For which motion was the Moon rotating?
2. For each type of motion, how many times did the Moon rotate during one revolution around Earth?
3. **Key Concept** How is the Moon actually rotating if the same side of the Moon is always facing Earth?

SCIENCE USE V. COMMON USE

phase

Science Use how the Moon or a planet is lit as seen from Earth

Common Use a part of something or a stage of development

Phases of the Moon

The Sun is always shining on half of the Moon, just as the Sun is always shining on half of Earth. However, as the Moon moves around Earth, usually only part of the Moon's near side is lit. *The portion of the Moon or a planet reflecting light as seen from Earth is called a* **phase**. As shown in **Figure 12**, the motion of the Moon around Earth causes the phase of the Moon to change. The sequence of phases is the lunar cycle. One lunar cycle takes 29.5 days or slightly more than four weeks to complete.

Key Concept Check What produces the phases of the Moon?

Waxing Phases

During the **waxing phases**, *more of the Moon's near side is lit each night.*

Week 1—First Quarter As the lunar cycle begins, a sliver of light can be seen on the Moon's western edge. Gradually the lit part becomes larger. By the end of the first week, the Moon is at its first quarter phase. In this phase, the Moon's entire western half is lit.

Week 2—Full Moon During the second week, more and more of the near side becomes lit. When the Moon's near side is completely lit, it is at the full moon phase.

Waning Phases

During the **waning phases**, *less of the Moon's near side is lit each night.* As seen from Earth, the lit part is now on the Moon's eastern side.

Week 3—Third Quarter During this week, the lit part of the Moon becomes smaller until only the eastern half of the Moon is lit. This is the third quarter phase.

Week 4—New Moon During this week, less and less of the near side is lit. When the Moon's near side is completely dark, it is at the new moon phase.

The Lunar Cycle

Figure 12 As the Moon revolves around Earth, the part of the Moon's near side that is lit changes. The figure below shows how the Moon looks at different places in its orbit.

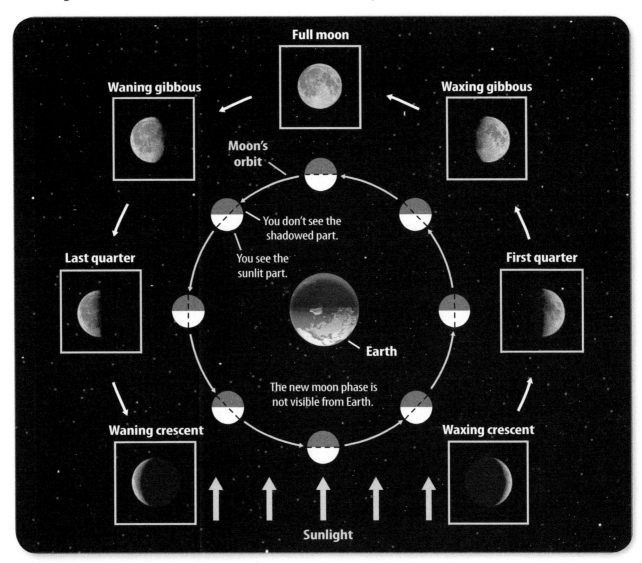

The Moon at Midnight

The Moon's motion around Earth causes the Moon to rise, on average, about 50 minutes later each day. The figure below shows how the Moon looks at midnight during three phases of the lunar cycle.

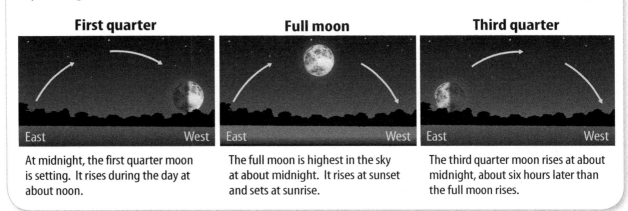

At midnight, the first quarter moon is setting. It rises during the day at about noon.

The full moon is highest in the sky at about midnight. It rises at sunset and sets at sunrise.

The third quarter moon rises at about midnight, about six hours later than the full moon rises.

Lesson 2 Review

Visual Summary

According to the giant impact hypothesis, a large object collided with Earth about 4.5 billion years ago to form the Moon.

Features like maria, craters, and highlands formed on the Moon's surface early in its history.

The Moon's phases change in a regular pattern during the Moon's lunar cycle.

FOLDABLES

Use your lesson Foldable to review the lesson. Save your Foldable for the project at the end of the chapter.

What do you think NOW?

You first read the statements below at the beginning of the chapter.

3. The Moon was once a planet that orbited the Sun between Earth and Mars.

4. Earth's shadow causes the changing appearance of the Moon.

Did you change your mind about whether you agree or disagree with the statements? Rewrite any false statements to make them true.

Use Vocabulary

1. The lit part of the Moon as viewed from Earth is a(n) _____.

2. For the first half of the lunar cycle, the lit part of the Moon's near side is _____.

3. For the second half of the lunar cycle, the lit part of the Moon's near side is _____.

Understand Key Concepts

4. Which phase occurs when the Moon is between the Sun and Earth?
   A. first quarter   C. new moon
   B. full moon   D. third quarter

5. **Reason** Why does the Moon have phases?

Interpret Graphics

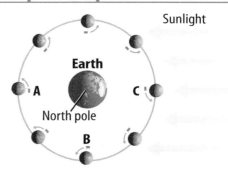

6. **Draw** how the Moon looks from Earth when it is at positions A, B, and C in the diagram above.

7. **Organize Information** Copy and fill in the table below with details about the lunar surface.

Crater	
Ray	
Maria	
Highland	

Critical Thinking

8. **Reflect** Imagine the Moon rotates twice in the same amount of time the Moon orbits Earth once. Would you be able to see the Moon's far side from Earth?

740 Chapter 20
EVALUATE

SCIENCE & SOCIETY

Return to the Moon

Exploring Earth's Moon is a step toward exploring other planets and building outposts in space.

The United States undertook a series of human spaceflight missions from 1961–1975 called the Apollo program. The goal of the program was to land humans on the Moon and bring them safely back to Earth. Six of the missions reached this goal. The Apollo program was a huge success, but it was just the beginning.

NASA began another space program that had a goal to return astronauts to the Moon to live and work. However, before that could happen, scientists needed to know more about conditions on the Moon and what materials are available there.

Collecting data was the first step. In 2009, NASA launched the *Lunar Reconnaissance Orbiter (LRO)* spacecraft. The *LRO* spent a year orbiting the Moon's two poles. It collected detailed data that scientists can use to make maps of the Moon's features and resources, such as deep craters that formed on the Moon when comets and asteroids slammed into it billions of years ago. Some scientists predicted that these deep craters contain frozen water.

One of the instruments launched with the *LRO* was the *Lunar Crater Observation and Sensing Satellite (LCROSS)*. *LCROSS* observations confirmed the scientists' predictions that water exists on the Moon. A rocket launched from *LCROSS* impacted the Cabeus crater near the Moon's south pole. The material that was ejected after the rocket's impact included water.

NASA's goal of returning astronauts to the Moon was delayed, and their missions now focus on exploring Mars instead. But the discoveries made on the Moon will help scientists develop future missions that could take humans farther into the solar system.

Apollo SPACE PROGRAM

The Apollo Space Program included 17 missions. Here are some milestones:

January 27 1967
Apollo 1 Fire killed all three astronauts on board during a launch simulation for the first piloted flight to the Moon.

December 21–27 1968
Apollo 8 First manned spacecraft orbits the Moon.

July 16–24 1969
Apollo 11 First humans, Neil Armstrong and Buzz Aldrin, walk on the Moon.

July 1971
Apollo 15 Astronauts drive the first rover on the Moon.

December 7–19 1972
Apollo 17 The first phase of human exploration of the Moon ended with this last lunar landing mission.

It's Your Turn

BRAINSTORM As a group, brainstorm the different occupations that would be needed to successfully operate a base on the Moon or another planet. Discuss the tasks that a person would perform in each occupation.

Lesson 3

Eclipses and Tides

Reading Guide

Key Concepts
ESSENTIAL QUESTIONS

- What is a solar eclipse?
- What is a lunar eclipse?
- How do the Moon and the Sun affect Earth's oceans?

Vocabulary
umbra p. 743
penumbra p. 743
solar eclipse p. 744
lunar eclipse p. 746
tide p. 747

Multilingual eGlossary

Video

- BrainPOP®
- Science Video

Inquiry What is this dark spot?

A NASA satellite took this photo as it orbited around Earth. An eclipse caused the shadow that you see. Do you know what kind of eclipse?

742 • Chapter 20
ENGAGE

Launch Lab

10 minutes

How do shadows change?

You can see a shadow when an object blocks a light source. What happens to an object's shadow when the object moves?

1. Read and complete a lab safety form.
2. Select an **object** provided by your teacher.
3. Shine a **flashlight** on the object, projecting its shadow on the wall.
4. While holding the flashlight in the same position, move the object closer to the wall—away from the light. Then, move the object toward the light. Record your observations in your Science Journal.

Think About This

1. Compare and contrast the shadows created in each situation. Did the shadows have dark parts and light parts? Did these parts change?

2. **Key Concept** Imagine you look at the flashlight from behind your object, looking from the darkest and lightest parts of the object's shadow. How much of the flashlight could you see from each location?

Shadows—the Umbra and the Penumbra

A shadow results when one object blocks the light that another object emits or reflects. When a tree blocks light from the Sun, it casts a shadow. If you want to stand in the shadow of a tree, the tree must be in a line between you and the Sun.

If you go outside on a sunny day and look carefully at a shadow on the ground, you might notice that the edges of the shadow are not as dark as the rest of the shadow. Light from the Sun and other wide sources casts shadows with two distinct parts, as shown in **Figure 13**. *The* **umbra** *is the central, darker part of a shadow where light is totally blocked. The* **penumbra** *is the lighter part of a shadow where light is partially blocked.* If you stood within an object's penumbra, you would be able to see only part of the light source. If you stood within an object's umbra, you would not see the light source at all.

WORD ORIGIN

penumbra
from Latin *paene*, means "almost"; and *umbra*, means "shade, shadow"

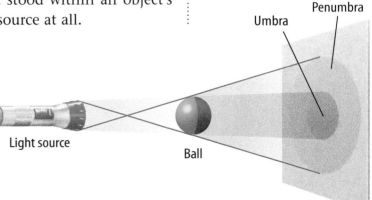

Figure 13 The shadow that a wide light source produces has two parts—the umbra and the penumbra. The light source cannot be seen from within the umbra. The light source can be partially seen from within the penumbra.

Inquiry MiniLab

10 minutes

What does the Moon's shadow look like?

Like every shadow cast by a wide light source, the Moon's shadow has two parts.

1. Read and complete a lab safety form.
2. Working with a partner, use a **pencil** to connect two **foam balls**. One ball should be one-fourth the size of the other.
3. While one person holds the balls, the other should stand 1 m away and shine a **flashlight** or **desk lamp** on the balls. The balls and light should be in a direct line, with the smallest ball closest to the light.
4. Sketch and describe your observations in your Science Journal.

Analyze and Conclude

1. **Key Concept** Explain the relationship between the two types of shadows and solar eclipses.

Solar Eclipses

As the Sun shines on the Moon, the Moon casts a shadow that extends out into space. Sometimes the Moon passes between Earth and the Sun. This can only happen during the new moon phase. When Earth, the Moon, and the Sun are lined up, the Moon casts a shadow on Earth's surface, as shown in **Figure 14.** You can see the Moon's shadow in the photo at the beginning of this lesson. *When the Moon's shadow appears on Earth's surface, a* **solar eclipse** *is occurring.*

 Key Concept Check Why does a solar eclipse occur only during a new moon?

As Earth rotates, the Moon's shadow moves along Earth's surface, as shown in **Figure 14.** The type of eclipse you see depends on whether you are in the path of the umbra or the penumbra. If you are outside the umbra and penumbra, you cannot see a solar eclipse at all.

Total Solar Eclipses

You can only see a total solar eclipse from within the Moon's umbra. During a total solar eclipse, the Moon appears to cover the Sun completely, as shown in **Figure 15** on the next page. Then, the sky becomes dark enough that you can see stars. A total solar eclipse lasts no longer than about 7 minutes.

Solar Eclipse

Figure 14 A solar eclipse occurs only when the Moon moves directly between Earth and the Sun. The Moon's shadow moves across Earth's surface.

Visual Check Why would a person in North America not see the solar eclipse shown here?

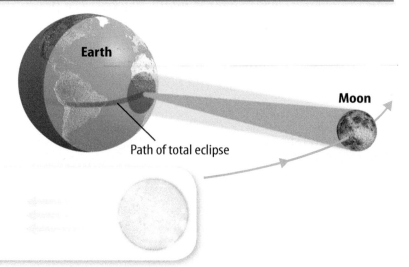

EXPLAIN

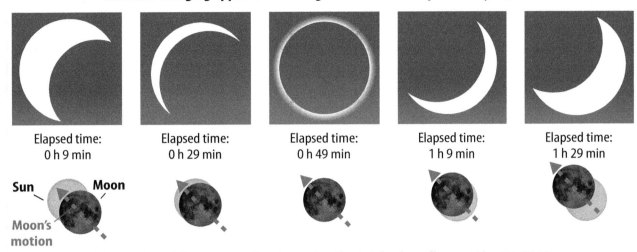

The sun's changing appearance during the total solar eclipse on May 20, 2012

Elapsed time: 0 h 9 min
Elapsed time: 0 h 29 min
Elapsed time: 0 h 49 min
Elapsed time: 1 h 9 min
Elapsed time: 1 h 29 min

Sun Moon
Moon's motion

The motion of the moon in the sky during the total solar eclipse on May 20, 2012

Partial Solar Eclipses

You can only see a total solar eclipse from within the Moon's umbra, but you can see a partial solar eclipse from within the Moon's much larger penumbra. The stages of a partial solar eclipse are similar to the stages of a total solar eclipse, except that the Moon never completely covers the Sun.

Why don't solar eclipses occur every month?

Solar eclipses only can occur during a new moon, when Earth and the Sun are on opposite sides of the Moon. However, solar eclipses do not occur during every new moon phase. **Figure 16** shows why. The Moon's orbit is tilted slightly compared to Earth's orbit. As a result, during most new moons, Earth is either above or below the Moon's shadow. However, every so often the Moon is in a line between the Sun and Earth. Then the Moon's shadow passes over Earth and a solar eclipse occurs.

Figure 15 This sequence of photographs shows how the Sun's appearance changed during a total solar eclipse in 2006.

Visual Check How much time elapsed from the start to the finish of this sequence?

Figure 16 A solar eclipse occurs only when the Moon crosses Earth's orbit and is in a direct line between Earth and the Sun.

The Moon's Tilted Orbit

Concepts in Motion Animation

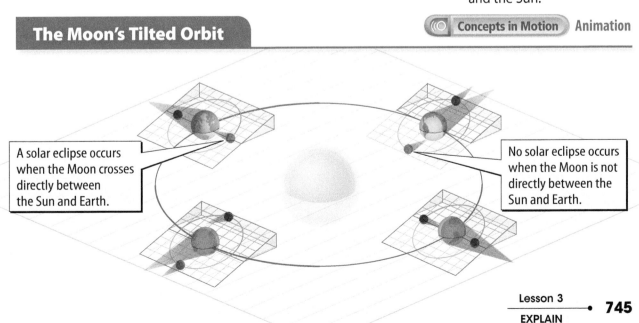

A solar eclipse occurs when the Moon crosses directly between the Sun and Earth.

No solar eclipse occurs when the Moon is not directly between the Sun and Earth.

Lunar Eclipse

Figure 17 A lunar eclipse occurs when the Moon moves through Earth's shadow.

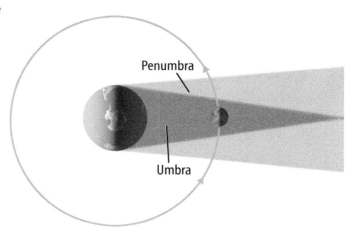

Visual Check Why would more people be able to see a lunar eclipse than a solar eclipse?

Lunar Eclipses

Just like the Moon, Earth casts a shadow into space. As the Moon revolves around Earth, it sometimes moves into Earth's shadow, as shown in **Figure 17.** A **lunar eclipse** *occurs when the Moon moves into Earth's shadow.* Then Earth is in a line between the Sun and the Moon. This means that a lunar eclipse can occur only during the full moon phase.

Like the Moon's shadow, Earth's shadow has an umbra and a penumbra. Different types of lunar eclipses occur depending on which part of Earth's shadow the Moon moves through. Unlike solar eclipses, you can see any lunar eclipse from any location on the side of Earth facing the Moon.

Key Concept Check When can a lunar eclipse occur?

Total Lunar Eclipses

When the entire Moon moves through Earth's umbra, a total lunar eclipse occurs. **Figure 18** on the next page shows how the Moon's appearance changes during a total lunar eclipse. The Moon's appearance changes as it gradually moves into Earth's penumbra, then into Earth's umbra, back into Earth's penumbra, and then out of Earth's shadow entirely.

You can still see the Moon even when it is completely within Earth's umbra. Although Earth blocks most of the Sun's rays, Earth's atmosphere deflects some sunlight into Earth's umbra. This is also why you can often see the unlit portion of the Moon on a clear night. Astronomers often call this Earthshine. This reflected light has a reddish color and gives the Moon a reddish tint during a total lunar eclipse.

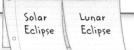

FOLDABLES
Make a two-tab book from a sheet of notebook paper. Label the tabs *Solar Eclipse* and *Lunar Eclipse*. Use it to organize your notes on eclipses.

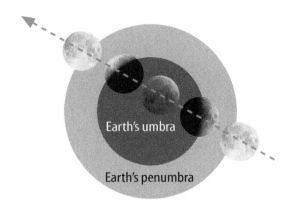

Figure 18 If the entire Moon passes through Earth's umbra, the Moon gradually darkens until a dark shadow covers it completely.

✓ **Visual Check** How would a total lunar eclipse look different from a total solar eclipse?

Partial Lunar Eclipses

When only part of the Moon passes through Earth's umbra, a partial lunar eclipse occurs. The stages of a partial lunar eclipse are similar to those of a total lunar eclipse, shown in **Figure 18,** except the Moon is never completely covered by Earth's umbra. The part of the Moon in Earth's penumbra appears only slightly darker, while the part of the Moon in Earth's umbra appears much darker.

Why don't lunar eclipses occur every month?

Lunar eclipses can only occur during a full moon phase, when the Moon and the Sun are on opposite sides of Earth. However, lunar eclipses do not occur during every full moon because of the tilt of the Moon's orbit with respect to Earth's orbit. During most full moons, the Moon is slightly above or slightly below Earth's penumbra.

Tides

The positions of the Moon and the Sun also affect Earth's oceans. If you have spent time near an ocean, you might have seen how the ocean's height, or sea level, rises and falls twice each day. *A* **tide** *is the daily rise and fall of sea level.* Examples of tides are shown in **Figure 19.** It is primarily the Moon's gravity that causes Earth's oceans to rise and fall twice each day.

Figure 19 In the Bay of Fundy, high tides can be more than 10 m higher than low tides.

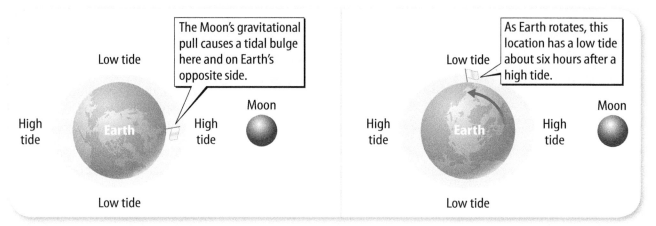

Figure 20 🔑 In this view down on Earth's North Pole, the flag moves into a tidal bulge as Earth rotates. A coastal area has a high tide about once every 12 hours.

The Moon's Effect on Earth's Tides

The difference in the strength of the Moon's gravity on opposite sides of Earth causes Earth's tides. The Moon's gravity is slightly stronger on the side of Earth closer to the Moon and slightly weaker on the side of Earth opposite the Moon. These differences cause tidal bulges in the oceans on opposite sides of Earth, shown in **Figure 20**. High tides occur at the tidal bulges, and low tides occur between them.

The Sun's Effect on Earth's Tides

Because the Sun is so far away from Earth, its effect on tides is about half that of the Moon. **Figure 21** shows how the positions of the Sun and the Moon affect Earth's tides.

Spring Tides During the full moon and new moon phases, spring tides occur. This is when the Sun's and the Moon's gravitational effects combine and produce higher high tides and lower low tides.

Neap Tides A week after a spring tide, a neap tide occurs. Then the Sun, Earth, and the Moon form a right angle. When this happens, the Sun's effect on tides reduces the Moon's effect. High tides are lower and low tides are higher at neap tides.

🔑 **Key Concept Check** Why is the Sun's effect on tides less than the Moon's effect?

Figure 21 A spring tide occurs when the Sun, Earth, and the Moon are in a line. A neap tide occurs when the Sun and the Moon form a right angle with Earth.

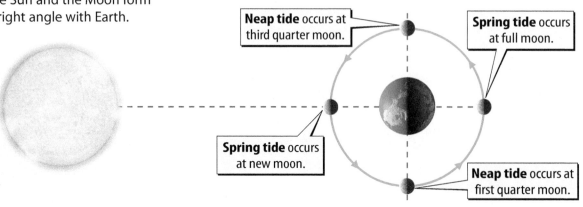

Lesson 3 Review

Assessment · Online Quiz

Visual Summary

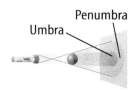

Shadows from a wide light source have two distinct parts.

The Moon's shadow produces solar eclipses. Earth's shadow produces lunar eclipses.

The positions of the Moon and the Sun in relation to Earth cause gravitational differences that produce tides.

FOLDABLES

Use your lesson Foldable to review the lesson. Save your Foldable for the project at the end of the chapter.

What do you think NOW?

You first read the statements below at the beginning of the chapter.

5. A solar eclipse happens when Earth moves between the Moon and the Sun.

6. The gravitational pull of the Moon and the Sun on Earth's oceans causes tides.

Did you change your mind about whether you agree or disagree with the statements? Rewrite any false statements to make them true.

Use Vocabulary

1 **Distinguish** between an umbra and a penumbra.

2 **Use the term** *tide* in a sentence.

3 The Moon turns a reddish color during a total _____ eclipse.

Understand Key Concepts

4 **Summarize** the effect of the Sun on Earth's tides.

5 **Illustrate** the positions of the Sun, Earth, and the Moon during a solar eclipse and during a lunar eclipse.

6 **Contrast** a total lunar eclipse with a partial lunar eclipse.

7 Which could occur during a total solar eclipse?
  A. first quarter moon   C. neap tide
  B. full moon   D. spring tide

Interpret Graphics

8 **Conclude** What type of eclipse does the figure above illustrate?

9 **Categorize Information** Copy and fill in the graphic organizer below to identify two bodies that affect Earth's tides.

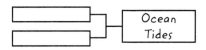

Critical Thinking

10 **Compose** a short story about a person long ago viewing a total solar eclipse.

11 **Research** ways to view a solar eclipse safely. Summarize your findings here.

Lesson 3 · **749**
EVALUATE

Inquiry Lab

35 minutes

Phases of the Moon

Materials

foam ball

pencil

lamp

stool

Safety

The Moon appears slightly different every night of its 29.5-day lunar cycle. The Moon's appearance changes as Earth and the Moon move. Depending on where the Moon is in relation to Earth and the Sun, observers on Earth see only part of the light the Moon reflects from the Sun.

Question

How do the positions of the Sun, the Moon, and Earth cause the phases of the Moon?

Procedure

1. Read and complete a lab safety form.
2. Hold a foam ball that represents the Moon. Make a handle for the ball by inserting a pencil about two inches into the ball. Your partner will represent an observer on Earth. Have your partner sit on a stool and record observations during the activity.
3. Place a lamp on a desk or other flat surface. Remove the shade from the lamp. The lamp represents the Sun.
4. Turn on the lamp and darken the lights in the room.
   ⚠ *Do not touch the bulb or look directly at it after the lamp is turned on.*
5. Position the Earth observer's stool about 1 m from the Sun. Position the Moon 0.5–1 m from the observer so that the Sun, Earth, and the Moon are in a line. The student holding the Moon holds the Moon so it is completely illuminated on one half. The observer records the phase and what the phase looks like in a data table.
6. Move the Moon clockwise about one-eighth of the way around its "orbit" of Earth. The observer swivels on the stool to face the Moon and records the phase.
7. Continue the Moon's orbit until the Earth observer has recorded all the Moon's phases.

8. Return to your positions as the Moon and Earth observer. Choose a part in the Moon's orbit that you did not model. Predict what the Moon would look like in that position, and check if your prediction is correct.

Analyze and Conclude

9. **Explain** Use your observations to explain how the positions of the Sun, the Moon, and Earth produce the different phases of the Moon.

10. **The Big Idea** Why is half of the Moon always lit? Why do you usually see only part of the Moon's lit half?

11. **Draw Conclusions** Based on your observations, why is the Moon not visible from Earth during the new moon phase?

12. **Summarize** Which parts of your model were waxing phases? Which parts were waning phases?

13. **Think Critically** During which phases of the Moon can eclipses occur? Explain.

Lab Tips

- ☑ Make sure the observer's head does not cast a shadow on the Moon.
- ☑ The student holding the Moon should hold the pencil so that he or she always stands on the unlit side of the Moon.

Communicate Your Results

Create a poster of the results from your lab. Illustrate various positions of the Sun, the Moon, and Earth and draw the phase of the Moon for each. Include a statement of your hypothesis on the poster.

The Moon is not the only object in the sky that has phases when viewed from Earth. The planets Venus and Mercury also have phases. Research the phases of these planets and create a calendar that shows when the various phases of Venus and Mercury occur.

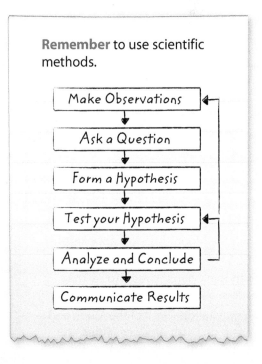

Remember to use scientific methods.

- Make Observations
- Ask a Question
- Form a Hypothesis
- Test your Hypothesis
- Analyze and Conclude
- Communicate Results

Lesson 3
751
EXTEND

Chapter 20 Study Guide

THE BIG IDEA: Earth's motion around the Sun causes seasons. The Moon's motion around Earth causes phases of the Moon. Earth and the Moon's motions together cause eclipses and ocean tides.

Key Concepts Summary

Lesson 1: Earth's Motion
- The gravitational pull of the Sun on Earth causes Earth to revolve around the Sun in a nearly circular **orbit**.
- Areas on Earth's curved surface become more tilted with respect to the direction of sunlight the farther you travel from the equator. This causes sunlight to spread out closer to the poles, making Earth colder at the poles and warmer at the equator.
- As Earth revolves around the Sun, the tilt of Earth's **rotation axis** produces changes in how sunlight spreads out over Earth's surface. These changes in the concentration of sunlight cause the seasons.

Vocabulary
orbit p. 726
revolution p. 726
rotation p. 727
rotation axis p. 727
solstice p. 731
equinox p. 731

Lesson 2: Earth's Moon

- The gravitational pull of Earth on the Moon makes the Moon revolve around Earth. The Moon rotates once as it makes one complete orbit around Earth.
- The lit part of the Moon that you can see from Earth—the Moon's **phase**—changes during the lunar cycle as the Moon revolves around Earth.

maria p. 736
phase p. 738
waxing phase p. 738
waning phase p. 738

Lesson 3: Eclipses and Tides
- When the Moon's shadow appears on Earth's surface, a **solar eclipse** occurs.
- When the Moon moves into Earth's shadow, a **lunar eclipse** occurs.
- The gravitational pull of the Moon and the Sun on Earth produces **tides**, the rise and fall of sea level that occurs twice each day.

umbra p. 743
penumbra p. 743
solar eclipse p. 744
lunar eclipse p. 746
tide p. 747

Study Guide

- Personal Tutor
- Vocabulary eGames
- Vocabulary eFlashcards

FOLDABLES Chapter Project

Assemble your Lesson Foldables as shown to make a Chapter Project. Use the project to review what you have learned in this chapter.

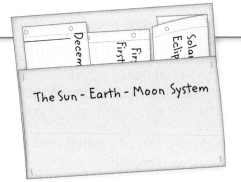

Use Vocabulary

Distinguish between the terms in the each of the following pairs.

1. revolution, orbit
2. rotation, rotation axis
3. solstice, equinox
4. waxing phases, waning phases
5. umbra, penumbra
6. solar eclipse, lunar eclipse
7. tide, phase

Link Vocabulary and Key Concepts

Concepts in Motion Interactive Concept Map

Copy this concept map, and then use vocabulary terms from the previous page to complete the concept map.

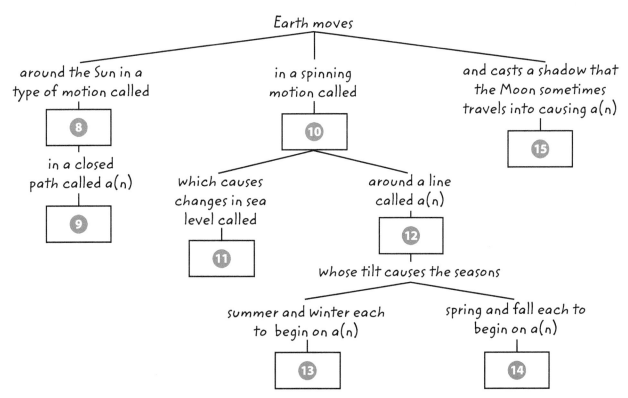

Chapter 20 Review

Understand Key Concepts

1. Which property of the Sun most affects the strength of gravitational attraction between the Sun and Earth?
   A. mass
   B. radius
   C. shape
   D. temperature

2. Which would be different if Earth rotated from east to west but at the same rate?
   A. the amount of energy striking Earth
   B. the days on which solstices occur
   C. the direction of the Sun's apparent motion across the sky
   D. the number of hours in a day

3. In the image below, which season is the northern hemisphere experiencing?

   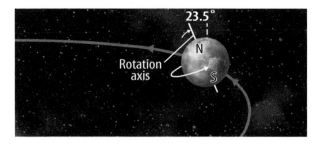

   A. fall
   B. spring
   C. summer
   D. winter

4. Which best explains why Earth is colder at the poles than at the equator?
   A. Earth is farther from the Sun at the poles than at the equator.
   B. Earth's orbit is not a perfect circle.
   C. Earth's rotation axis is tilted.
   D. Earth's surface is more tilted at the poles than at the equator.

5. How are the revolutions of the Moon and Earth alike?
   A. Both are produced by gravity.
   B. Both are revolutions around the Sun.
   C. Both orbits are the same size.
   D. Both take the same amount of time.

6. Which moon phase occurs about one week after a new moon?
   A. another new moon
   B. first quarter moon
   C. full moon
   D. third quarter moon

7. Why is the same side of the Moon always visible from Earth?
   A. The Moon does not revolve around Earth.
   B. The Moon does not rotate.
   C. The Moon makes exactly one rotation for each revolution around Earth.
   D. The Moon's rotation axis is not tilted.

8. About how often do spring tides occur?
   A. once each month
   B. once each year
   C. twice each month
   D. twice each year

9. If a coastal area has a high tide at 7:00 A.M., at about what time will the next low tide occur?
   A. 11:00 A.M.
   B. 1:00 P.M.
   C. 3:00 P.M.
   D. 7:00 P.M.

10. Which type of eclipse would a person standing at point X in the diagram below see?

    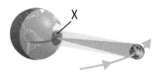

    A. partial lunar eclipse
    B. partial solar eclipse
    C. total lunar eclipse
    D. total solar eclipse

Chapter Review

Assessment
Online Test Practice

Critical Thinking

11 Outline the ways Earth moves and how each affects Earth.

12 Create a poster that illustrates and describes the relationship between Earth's tilt and the seasons.

13 Contrast Why can you see phases of the Moon but not phases of the Sun?

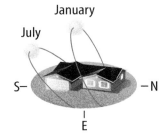

14 Interpret Graphics The figure above shows the Sun's position in the sky at noon in January and July. Is the house located in the northern hemisphere or the southern hemisphere? Explain.

15 Illustrate Make a diagram of the Moon's orbit and phases. Include labels and explanations with your drawing.

16 Differentiate between a total solar eclipse and a partial solar eclipse.

17 Generalize the reason that solar and lunar eclipses do not occur every month.

18 Role Play Write and present a play with several classmates that explains the causes and types of tides.

Writing in Science

19 Survey a group of at least ten people to determine how many know the cause of Earth's seasons. Write a summary of your results, including a main idea, supporting details, and a concluding sentence.

REVIEW THE BIG IDEA

20 At the South Pole, the Sun does not appear in the sky for six months out of the year. When does this happen? What is happening at the North Pole during these months? Explain why Earth's poles receive so little solar energy.

21 A solar eclipse, shown in the time-lapse photo below, is one phenomenon that the motions of Earth and the Moon produce. What other phenomena do the motions of Earth and the Moon produce?

Math Skills

Review Math Practice

Convert Units

22 When the Moon is 384,000 km from Earth, how far is the Moon from Earth in miles?

23 If you travel 205 mi on a train from Washington D.C. to New York City, how many kilometers do you travel on the train?

24 The nearest star other than the Sun is about 40 trillion km away. About how many miles away is the nearest star other than the Sun?

Standardized Test Practice

Record your answers on the answer sheet provided by your teacher or on a sheet of paper.

Multiple Choice

1. Which is the movement of one object around another object in space?
   A axis
   B orbit
   C revolution
   D rotation

Use the diagram below to answer question 2.

Time 1

Time 2

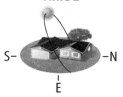

2. What happens between times *1* and *2* in the diagram above?
   A Days grow shorter and shorter.
   B The season changes from fall to winter.
   C The region begins to point away from the Sun.
   D The region gradually receives more solar energy.

3. How many times larger is the Sun's diameter than Earth's diameter?
   A about 10 times larger
   B about 100 times larger
   C about 1,000 times larger
   D about 10,000 times larger

4. Which diagram illustrates the Moon's third quarter phase?

   A

   B

   C

   D

5. Which accurately describes Earth's position and orientation during summer in the northern hemisphere?
   A Earth is at its closest point to the Sun.
   B Earth's hemispheres receive equal amounts of solar energy.
   C The north end of Earth's rotational axis leans toward the Sun.
   D The Sun emits a greater amount of light and heat energy.

6. Which are large, dark lunar areas formed by cooled lava?
   A craters
   B highlands
   C maria
   D rays

7. During one lunar cycle, the Moon
   A completes its east-to-west path across the sky exactly once.
   B completes its entire sequence of phases.
   C progresses only from the new moon phase to the full moon phase.
   D revolves around Earth twice.

Standardized Test Practice

Use the diagram below to answer question 8.

8 What does the flag in the diagram above represent?
 A high tide
 B low tide
 C neap tide
 D spring tide

9 During which lunar phase might a solar eclipse occur?
 A first quarter moon
 B full moon
 C new moon
 D third quarter moon

10 Which does the entire Moon pass through during a partial lunar eclipse?
 A Earth's penumbra
 B Earth's umbra
 C the Moon's penumbra
 D the Moon's umbra

Constructed Response

Use the diagram below to answer questions 11 and 12.

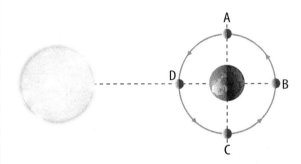

11 Where are neap tides indicated in the above diagram? What causes neap tides? What happens during a neap tide?

12 Where are spring tides indicated in the above diagram? What causes spring tides? What happens during a spring tide?

13 How would Earth's climate be different if its rotational axis were not tilted?

14 Why can we see only one side of the Moon from Earth? What is the name given to this side of the Moon?

15 What is a lunar phase? How do waxing and waning phases differ?

16 Why don't solar eclipses occur monthly?

NEED EXTRA HELP?																
If You Missed Question...	1	2	3	4	5	6	7	8	9	10	11	12	13	14	15	16
Go to Lesson...	1	1	1	2	1	2	2	3	3	3	3	3	1	2	2	3

Chapter 21

The Solar System

 What kinds of objects are in the solar system?

 One, Two, or Three Planets?

This photo, taken by the Cassini spacecraft, shows part of Saturn's rings and two of its moons. Saturn is a planet that orbits the Sun. The moons, tiny Epimetheus and much larger Titan, orbit Saturn. Besides planets and moons, many other objects are in the solar system.

- How would you describe a planet such as Saturn?
- How do astronomers classify the objects they discover?
- What types of objects do you think make up the solar system?

Get Ready to Read

What do you think?

Before you read, decide if you agree or disagree with each of these statements. As you read this chapter, see if you change your mind about any of the statements.

1. Astronomers measure distances between space objects using astronomical units.
2. Gravitational force keeps planets in orbit around the Sun.
3. Earth is the only inner planet that has a moon.
4. Venus is the hottest planet in the solar system.
5. The outer planets also are called the gas giants.
6. The atmospheres of Saturn and Jupiter are mainly water vapor.
7. Asteroids and comets are mainly rock and ice.
8. A meteoroid is a meteor that strikes Earth.

 Your one-stop online resource

connectED.mcgraw-hill.com

  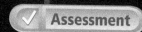

 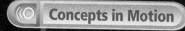

 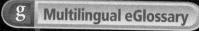

Lesson 1

Reading Guide

Key Concepts
ESSENTIAL QUESTIONS

- How are the inner planets different from the outer planets?
- What is an astronomical unit and why is it used?
- What is the shape of a planet's orbit?

Vocabulary
asteroid p. 763
comet p. 763
astronomical unit p. 764
period of revolution p. 764
period of rotation p. 764

 Multilingual eGlossary

 Video

- BrainPOP®
- Science Video

The Structure of the Solar System

Inquiry Are these stars?

Did you know that shooting stars are not actually stars? The bright streaks are small, rocky particles burning up as they enter Earth's atmosphere. These particles are part of the solar system and are often associated with comets.

Inquiry Launch Lab

10 minutes

How do you know which distance unit to use?

You can use different units to measure distance. For example, millimeters might be used to measure the length of a bolt, and kilometers might be used to measure the distance between cities. In this lab, you will investigate why some units are easier to use than others for certain measurements.

1. Read and complete a lab safety form.
2. Use a **centimeter ruler** to measure the length of a **pencil** and the thickness of this **book.** Record the distances in your Science Journal.
3. Use the centimeter ruler to measure the width of your classroom. Then measure the width of the room using a **meterstick.** Record the distances in your Science Journal.

Think About This

1. Why are meters easier to use than centimeters for measuring the classroom?
2. **Key Concept** Why do you think astronomers might need a unit larger than a kilometer to measure distances in the solar system?

What is the solar system?

Have you ever made a wish on a star? If so, you might have wished on a planet instead of a star. Sometimes, as shown in **Figure 1,** the first starlike object you see at night is not a star at all. It's Venus, the planet closest to Earth.

It's hard to tell the difference between planets and stars in the night sky because they all appear as tiny lights. Thousands of years ago, observers noticed that a few of these tiny lights moved, but others did not. The ancient Greeks called these objects planets, which means "wanderers." Astronomers now know that the planets do not wander about the sky; the planets move around the Sun. The Sun and the group of objects that move around it make up the solar system.

When you look at the night sky, a few of the tiny lights that you can see are part of our solar system. Almost all of the other specks of light are stars. They are much farther away than any objects in our solar system. Astronomers have discovered that some of those stars also have planets moving around them.

Reading Check What object do the planets in the solar system move around?

Figure 1 When looking at the night sky, you will likely see stars and planets. In the photo below, the planet Venus is the bright object seen above the Moon.

Objects in the Solar System

Ancient observers looking at the night sky saw many stars but only five planets—Mercury, Venus, Mars, Jupiter, and Saturn. The invention of the telescope in the 1600s led to the discovery of additional planets and many other space objects.

The Sun

The largest object in the solar system is the Sun, a *star*. Its diameter is about 1.4 million km—ten times the diameter of the largest planet, Jupiter. The Sun is made mostly of hydrogen gas. Its mass makes up about 99 percent of the entire solar system's mass.

Inside the Sun, a process called nuclear fusion produces an enormous amount of energy. The Sun emits some of this energy as light. The light from the Sun shines on all of the planets every day. The Sun also applies gravitational forces to objects in the solar system. Gravitational forces cause the planets and other objects to move around, or *orbit,* the Sun.

Objects That Orbit the Sun

Different types of objects orbit the Sun. These objects include planets, dwarf planets, asteroids, and comets. Unlike the Sun, these objects don't emit light but only reflect the Sun's light.

Planets Astronomers classify some objects that orbit the Sun as planets, as shown in **Figure 2**. An object is a planet only if it orbits the Sun and has a nearly spherical shape. Also, the mass of a planet must be much larger than the total mass of all other objects whose orbits are close by. The solar system has eight objects classified as planets.

Reading Check What is a planet?

SCIENCE USE V. COMMON USE

star
Science Use an object in space made of gases in which nuclear fusion reactions occur that emit energy
Common Use a shape that usually has five or six points around a common center

REVIEW VOCABULARY

orbit
(noun) the path an object follows as it moves around another object
(verb) to move around another object

Figure 2 The orbits of the inner and outer planets are shown to scale. The Sun and the planets are not to scale. The outer planets are much larger than the inner planets.

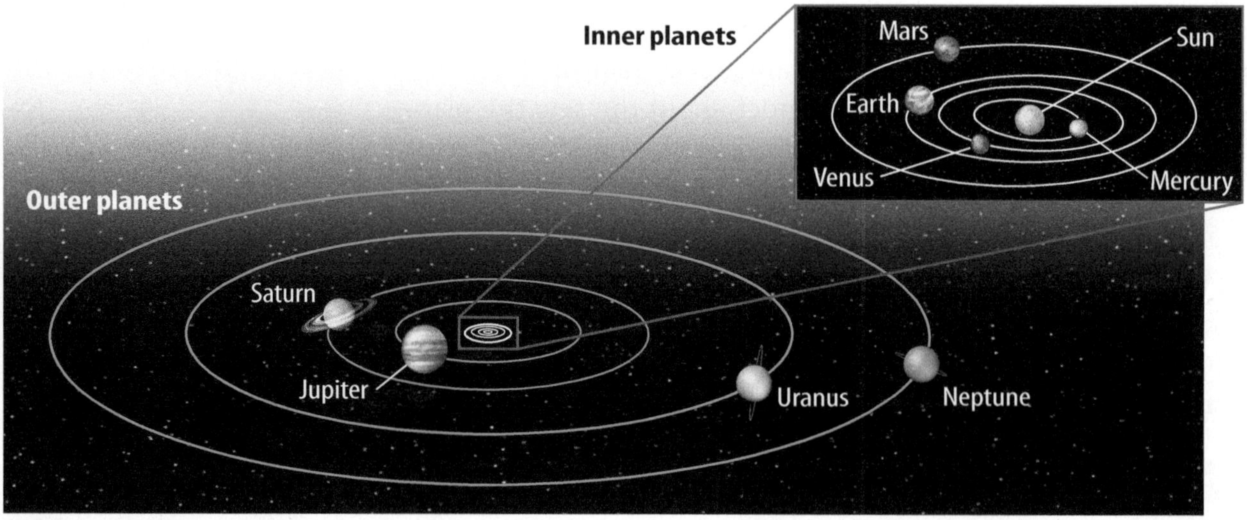

Inner planets: Mars, Sun, Earth, Venus, Mercury

Outer planets: Saturn, Jupiter, Uranus, Neptune

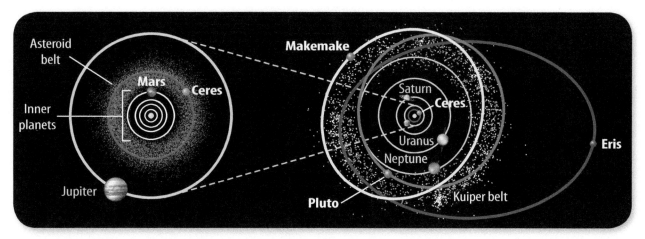

Inner Planets and Outer Planets As shown in **Figure 2**, the four planets closest to the Sun are the inner planets. The inner planets are Mercury, Venus, Earth, and Mars. These planets are made mainly of solid rocky materials. The four planets farthest from the Sun are the outer planets. The outer planets are Jupiter, Saturn, Uranus (YOOR uh nus), and Neptune. These planets are made mainly of ice and gases such as hydrogen and helium. The outer planets are much larger than Earth and are sometimes called gas giants.

 Key Concept Check Describe how the inner planets differ from the outer planets.

Dwarf Planets Scientists classify some objects in the solar system as dwarf planets. A dwarf planet is a spherical object that orbits the Sun. It is not a moon of another planet and is in a region of the solar system where there are many objects orbiting near it. But, unlike a planet, a dwarf planet does not have more mass than objects in nearby orbits. **Figure 3** shows the locations of the dwarf planets Ceres (SIHR eez), Eris (IHR is), Pluto, and Makemake (MAH kay MAH kay). Dwarf planets are made of rock and ice and are much smaller than Earth.

Asteroids *Millions of small, rocky objects called* **asteroids** *orbit the Sun in the asteroid belt between the orbits of Mars and Jupiter.* The asteroid belt is shown in **Figure 3**. Asteroids range in size from less than a meter to several hundred kilometers in length. Unlike planets and dwarf planets, asteroids, such as the one shown in **Figure 4**, usually are not spherical.

Comets You might have seen a picture of a comet with a long, glowing tail. *A* **comet** *is made of gas, dust, and ice and moves around the Sun in an oval-shaped orbit.* Comets come from the outer parts of the solar system. There might be 1 trillion comets orbiting the Sun. You will read more about comets, asteroids, and dwarf planets in Lesson 4.

▲ **Figure 3** Ceres, a dwarf planet, orbits the Sun as planets do. The orbit of Ceres is in the asteroid belt between Mars and Jupiter.

Visual Check Which dwarf planet is farthest from the Sun?

Animation

WORD ORIGIN
asteroid
from Greek *asteroeides*, means "resembling a star"

Figure 4 The asteroid Gaspra orbits the Sun in the asteroid belt. Its odd shape is about 19 km long and 11 km wide. ▼

The Astronomical Unit

On Earth, distances are often measured in meters (m) or kilometers (km). Objects in the solar system, however, are so far apart that astronomers use a larger distance unit. *An astronomical unit (AU) is the average distance from Earth to the Sun—about 150 million km.* Table 1 lists each planet's average distance from the Sun in km and AU.

 Key Concept Check Define what an astronomical unit is and explain why it is used.

Table 1 Because the distances of the planets from the Sun are so large, it is easier to express these distances using astronomical units rather than kilometers.

Concepts in Motion
Interactive Table

Table 1 Average Distance of the Planets from the Sun

Planet	Average Distance (km)	Average Distance (AU)
Mercury	57,910,000	0.39
Venus	108,210,000	0.72
Earth	149,600,000	1.00
Mars	227,920,000	1.52
Jupiter	778,570,000	5.20
Saturn	1,433,530,000	9.58
Uranus	2,872,460,000	19.20
Neptune	4,495,060,000	30.05

The Motion of the Planets

Have you ever swung a ball on the end of a string in a circle over your head? In some ways, the motion of a planet around the Sun is like the motion of that ball. As shown in Figure 5 on the next page, the Sun's gravitational force pulls each planet toward the Sun. This force is similar to the pull of the string that keeps the ball moving in a circle. The Sun's gravitational force pulls on each planet and keeps it moving along a curved path around the Sun.

 Reading Check What causes planets to orbit the Sun?

Revolution and Rotation

Objects in the solar system move in two ways. They orbit, or revolve, around the Sun. *The time it takes an object to travel once around the Sun is its* **period of revolution.** Earth's period of revolution is one year. The objects also spin, or rotate, as they orbit the Sun. *The time it takes an object to complete one rotation is its* **period of rotation.** Earth has a period of rotation of one day.

FOLDABLES
Make a tri-fold book from a sheet of paper and label it as shown. Use it to summarize information about the types of objects that make up the solar system.

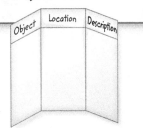

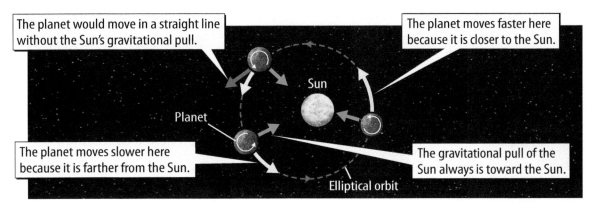

Planetary Orbits and Speeds

Unlike a ball swinging on the end of a string, planets do not move in circles. Instead, a planet's orbit is an ellipse—a stretched-out circle. Inside an ellipse are two special points, each called a focus. These focus points, or foci, determine the shape of the ellipse. The foci are equal distances from the center of the ellipse. As shown in **Figure 5**, the Sun is at one of the foci; the other foci is empty space. As a result, the distance between the planet and the Sun changes as the planet moves.

A planet's speed also changes as it orbits the Sun. The closer the planet is to the Sun, the faster it moves. This also means that planets farther from the Sun have longer periods of revolution. For example, Jupiter is more than five times farther from the Sun than Earth. Not surprisingly, Jupiter takes 12 times longer than Earth to revolve around the Sun.

Figure 5 Planets and other objects in the solar system revolve around the Sun because of its gravitational pull on them.

Review
Personal Tutor

 Key Concept Check Describe the shape of a planet's orbit.

Inquiry MiniLab
20 minutes

How can you model an elliptical orbit?
In this lab you will explore how the locations of foci affect the shape of an ellipse.

1. Read and complete a lab safety form.
2. Place a sheet of **paper** on a **corkboard.** Insert two **push pins** 8 cm apart in the center of the paper.
3. Use **scissors** to cut a 24-cm piece of **string.** Tie the ends of the string together.
4. Place the loop of string around the pins. Use a pencil to draw an ellipse as shown.
5. Measure the maximum width and length of the ellipse. Record the data in your Science Journal.
6. Move one of the push pins so that the pins are 5 cm apart. Repeat steps 4 and 5.

Analysis
1. **Compare and contrast** the two ellipses.
2.  **Key Concept** How are the shapes of the ellipses you drew similar to the orbits of the inner and outer planets?

Lesson 1 Review

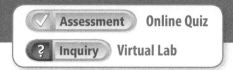

Visual Summary

The solar system contains the Sun, the inner planets, the outer planets, the dwarf planets, asteroids, and comets.

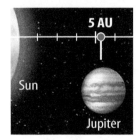

An astronomical unit (AU) is a unit of distance equal to about 150 million km.

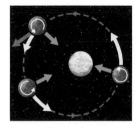

The speeds of the planets change as they move around the Sun in elliptical orbits.

Use your lesson Foldable to review the lesson. Save your Foldable for the project at the end of the chapter.

What do you think NOW?

You first read the statements below at the beginning of the chapter.

1. Astronomers measure distances between space objects using astronomical units.
2. Gravitational force keeps planets in orbit around the Sun.

Did you change your mind about whether you agree or disagree with the statements? Rewrite any false statements to make them true.

Use Vocabulary

1. **Compare and contrast** a period of revolution and a period of rotation.
2. **Define** *dwarf planet* in your own words.
3. **Distinguish** between an asteroid and a comet.

Understand Key Concepts

4. **Summarize** how and why planets orbit the Sun and how and why a planet's speed changes in orbit.
5. **Infer** why an astronomical unit is not used to measure distances on Earth.
6. Which distinguishes a dwarf planet from a planet?
   A. mass
   B. the object it revolves around
   C. shape
   D. type of orbit

Interpret Graphics

7. **Explain** what each arrow in the diagram represents.

8. **Take Notes** Copy the table below. List information about each object or group of objects in the solar system mentioned in the lesson. Add additional lines as needed.

Object	Description
Sun	
Planets	

Critical Thinking

9. **Evaluate** How would the speed of a planet be different if its orbit were a circle instead of an ellipse?

Meteors are pieces of a comet or an asteroid that heat up as they fall through Earth's atmosphere. Meteors that strike Earth are called meteorites.

History from Space

CAREERS in SCIENCE

AMERICAN MUSEUM OF NATURAL HISTORY

Meteorites give a peek back in time.

About 4.6 billion years ago, Earth and the other planets did not exist. In fact, there was no solar system. Instead, a large disk of gas and dust, known as the solar nebula, swirled around a forming Sun, as shown in the top picture to the right. How did the planets and other objects in the solar system form?

Denton Ebel is looking for the answer. He is a geologist at the American Museum of Natural History in New York City. Ebel explores the hypothesis that over millions of years, tiny particles in the solar nebula clumped together and formed the asteroids, comets, and planets that make up our solar system.

The solar nebula contained tiny particles called chondrules (KON drewls). They formed when the hot gas of the nebula condensed and solidified. Chondrules and other tiny particles collided and then accreted (uh KREET ed) or clumped together. This process eventually formed asteroids, comets, and planets. Some of the asteroids and comets have not changed much in over 4 billion years. Chondrite meteorites are pieces of asteroids that fell to Earth. The chondrules within the meteorites are the oldest solid material in our solar system.

For Ebel, chondrite meteorites contain information about the formation of the solar system. Did the materials in the meteorite form throughout the solar system and then accrete? Or did asteroids and comets form and accrete near the Sun, drift outward to where they are today, and then grow larger by accreting ice and dust? Ebel's research is helping to solve the mystery of how our solar system formed.

▲ Denton Ebel holds a meteorite that broke off the Vesta asteroid.

Accretion Hypothesis

According to the accretion hypothesis, the solar system formed in stages.

First there was a solar nebula. The Sun formed when gravity caused the nebula to collapse.

The rocky inner planets formed from accreted particles.

The gaseous outer planets formed as gas, ice, and dust condensed and accreted.

It's Your Turn

TIME LINE Work in groups. Learn more about the history of Earth from its formation until life began to appear. Create a time line showing major events. Present your time line to the class.

Lesson 1
EXTEND

Lesson 2

Reading Guide

Key Concepts
ESSENTIAL QUESTIONS

- How are the inner planets similar?
- Why is Venus hotter than Mercury?
- What kind of atmospheres do the inner planets have?

Vocabulary
terrestrial planet p. 769
greenhouse effect p. 771

 Multilingual eGlossary

 Video
What's Science Got to do With It?

The Inner Planets

Inquiry) Where is this?

This spectacular landscape is the surface of Mars, one of the inner planets. Other inner planets have similar rocky surfaces. It might surprise you to learn that there are planets in the solar system that have no solid surface on which to stand.

Inquiry Launch Lab

20 minutes

What affects the temperature on the inner planets?

Mercury and Venus are closer to the Sun than Earth. What determines the temperature on these planets? Let's find out.

1. Read and complete a lab safety form.
2. Insert a **thermometer** into a **clear 2-L plastic bottle.** Wrap **modeling clay** around the lid to hold the thermometer in the center of the bottle. Form an airtight seal with the clay.
3. Rest the bottle against the side of a **shoe box** in direct sunlight. Lay a second **thermometer** on top of the box next to the bottle so that the bulbs are at about the same height. The thermometer bulb should not touch the box. Secure the thermometer in place using **tape.**
4. Read the thermometers and record the temperatures in your Science Journal.
5. Wait 15 minutes and then read and record the temperature on each thermometer.

Think About This

1. How did the temperature of the two thermometers compare?
2. **Key Concept** What do you think caused the difference in temperature?

Planets Made of Rock

Imagine that you are walking outside. How would you describe the ground? You might say it is dusty or grassy. If you live near a lake or an ocean, you might say the ground is sandy or wet. But beneath the ground or lake or ocean is a layer of solid rock.

The inner planets—Mercury, Venus, Earth, and Mars—are also called terrestrial planets. **Terrestrial planets** *are the planets closest to the Sun, are made of rock and metal, and have solid outer layers.* Like Earth, the other inner planets also are made of rock and metallic materials and have a solid outer layer. However, as shown in **Figure 6,** the inner planets have different sizes, atmospheres, and surfaces.

WORD ORIGIN

terrestrial
from Latin *terrestris,* means "earthly"

Figure 6 The inner planets are roughly similar in size. Earth is about two and half times larger than Mercury. All inner planets have a solid outer layer.

INNER PLANETS

Mercury   Venus   Earth   Mars

Visual Check Which is the smallest inner planet?

Lesson 2
EXPLORE

Figure 7 The *Messenger* space probe flew by Mercury in 2008 and photographed the planet's cratered surface.

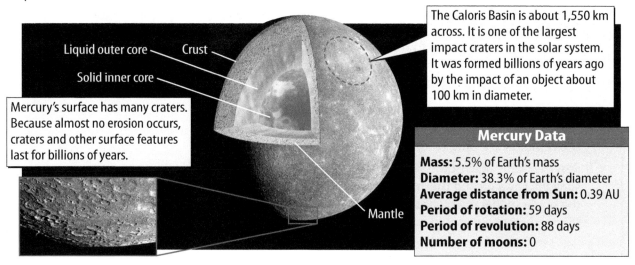

Mercury's surface has many craters. Because almost no erosion occurs, craters and other surface features last for billions of years.

The Caloris Basin is about 1,550 km across. It is one of the largest impact craters in the solar system. It was formed billions of years ago by the impact of an object about 100 km in diameter.

Mercury Data
Mass: 5.5% of Earth's mass
Diameter: 38.3% of Earth's diameter
Average distance from Sun: 0.39 AU
Period of rotation: 59 days
Period of revolution: 88 days
Number of moons: 0

FOLDABLES

Make a four-door book. Label each door with the name of an inner planet. Use the book to organize your notes on the inner planets.

Mercury

The smallest planet and the planet closest to the Sun is Mercury, shown in **Figure 7.** Mercury has no atmosphere. A planet has an atmosphere when its gravity is strong enough to hold gases close to its surface. The strength of a planet's gravity depends on the planet's mass. Because Mercury's mass is so small, its gravity is not strong enough to hold onto an atmosphere. Without an atmosphere there is no wind that moves energy from place to place across the planet's surface. This results in temperatures as high as 450°C on the side of Mercury facing the Sun and as cold as −170°C on the side facing away from the Sun.

Mercury's Surface

Impact craters, depressions formed by collisions with objects from space, cover the surface of Mercury. There are smooth plains of solidified lava from long-ago eruptions. There are also high cliffs that might have formed when the planet cooled quickly, causing the surface to wrinkle and crack. Without an atmosphere, almost no erosion occurs on Mercury's surface. As a result, features that formed billions of years ago have changed very little.

Mercury's Structure

The structures of the inner planets are similar. Like all inner planets, Mercury has a core made of iron and nickel. Surrounding the core is a layer called the mantle. The mantle is mainly made of silicon and oxygen. The crust is a thin, rocky layer above the mantle. Mercury's large core might have been formed by a collision with a large object during Mercury's formation.

Key Concept Check How are the inner planets similar?

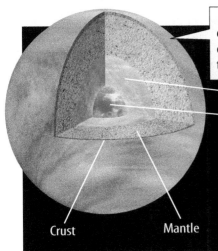

Venus's atmosphere traps energy. The greenhouse effect greatly increases the planet's temperature.

The *Magellan* orbiter used radar that can see through Venus's clouds to make images of the planet's surface.

Liquid outer core
Solid inner core
Crust
Mantle

This radar image shows a volcano on Venus's surface.

Venus Data
Mass: 81.5% of Earth's mass
Diameter: 95% of Earth's diameter
Average distance from Sun: 0.72 AU
Period of rotation: 244 days
Period of revolution: 225 days
Number of moons: 0

Venus

The second planet from the Sun is Venus, as shown in **Figure 8**. Venus is about the same size as Earth. It rotates so slowly that its period of rotation is longer than its period of revolution. This means that a day on Venus is longer than a year. Unlike most planets, Venus rotates from east to west. Several space probes have flown by or landed on Venus.

Venus's Atmosphere

The atmosphere of Venus is about 97 percent carbon dioxide. It is so dense that the atmospheric pressure on Venus is about 90 times greater than on Earth. Even though Venus has almost no water in its atmosphere or on its surface, a thick layer of clouds covers the planet. Unlike the clouds of water vapor on Earth, the clouds on Venus are made of acid.

The Greenhouse Effect on Venus

With an average temperature of about 460°C, Venus is the hottest planet in the solar system. The high temperatures are caused by the greenhouse effect. *The* **greenhouse effect** *occurs when a planet's atmosphere traps solar energy and causes the surface temperature to increase.* Carbon dioxide in Venus's atmosphere traps some of the solar energy that is absorbed and then emitted by the planet. This heats up the planet. Without the greenhouse effect, Venus would be almost 450°C cooler.

Key Concept Check Why is Venus hotter than Mercury?

Venus's Structure and Surface

Venus's internal structure, as shown in **Figure 8,** is similar to Earth's. Radar images show that more than 80 percent of Venus's surface is covered by solidified lava. Much of this lava might have been produced by volcanic eruptions that occurred about half a billion years ago.

Figure 8 Because a thick layer of clouds covers Venus, its surface has not been seen. Between 1990 and 1994, the *Magellan* space probe mapped the surface using radar.

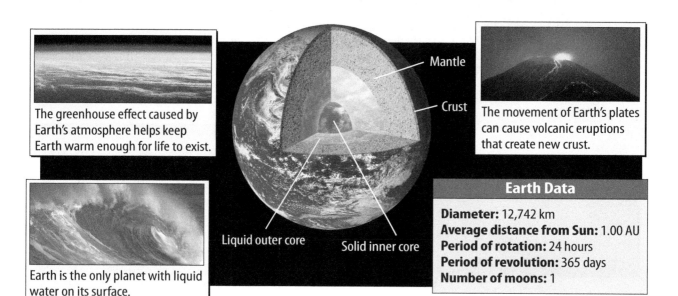

Figure 9 Earth has more water in its atmosphere and on its surface than the other inner planets. Earth's surface is younger than the surfaces of the other inner planets because new crust is constantly forming.

Inquiry MiniLab
20 minutes

How can you model the inner planets?

In this lab, you will use modeling clay to make scale models of the inner planets.

Planet	Actual Diameter (km)	Model Diameter (cm)
Mercury	4,879	
Venus	12,103	
Earth	12,756	8.0
Mars	6,792	

1. Use the data above for Earth to calculate in your Science Journal each model's diameter for the other three planets.
2. Use **modeling clay** to make a ball that represents the diameter of each planet. Check the diameter with a **centimeter ruler**.

Analyze Your Results

1. **Explain** how you converted actual diameters (km) to model diameters (cm).
2. **Key Concept** How do the inner planets compare? Which planets have approximately the same diameter?

Earth

Earth, shown in **Figure 9,** is the third planet from the Sun. Unlike Mercury and Venus, Earth has a moon.

Earth's Atmosphere

A mixture of gases and a small amount of water vapor make up most of Earth's atmosphere. They produce a greenhouse effect that increases Earth's average surface temperature. This effect and Earth's distance from the Sun warm Earth enough for large bodies of liquid water to exist. Earth's atmosphere also absorbs much of the Sun's radiation and protects the surface below. Earth's protective atmosphere, the presence of liquid water, and the planet's moderate temperature range support a variety of life.

Earth's Structure

As shown in **Figure 9,** Earth has a solid inner core surrounded by a liquid outer core. The mantle surrounds the liquid outer core. Above the mantle is Earth's crust. It is broken into large pieces, called plates, that constantly slide past, away from, or into each other. The crust is made mostly of oxygen and silicon and is constantly created and destroyed.

Reading Check Why is there life on Earth?

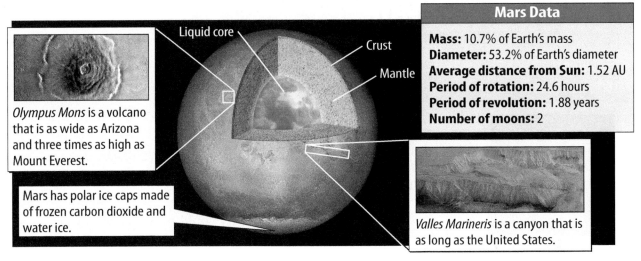

▲ Figure 10 Mars is a small, rocky planet with deep canyons and tall mountains.

Mars

The fourth planet from the Sun is Mars, shown in **Figure 10.** Mars is about half the size of Earth. It has two very small and irregularly shaped moons. These moons might be asteroids that were captured by Mars's gravity.

Many space probes have visited Mars. Most of them have searched for signs of water that might indicate the presence of living organisms. Images of Mars show features that might have been made by water, such as the gullies in **Figure 11.** So far no evidence of liquid water or life has been found.

Mars's Atmosphere

The atmosphere of Mars is about 95 percent carbon dioxide. It is thin and much less dense than Earth's atmosphere. Temperatures range from about −125°C at the poles to about 20°C at the equator during a martian summer. Winds on Mars sometimes produce great dust storms that last for months.

Mars's Surface

The reddish color of Mars is because its soil contains iron oxide, a compound in rust. Some of Mars's major surface features are shown in **Figure 10.** The enormous canyon Valles Marineris is about 4,000 km long. The Martian volcano Olympus Mons is the largest known mountain in the solar system. Mars also has polar ice caps made of frozen carbon dioxide and ice.

The southern hemisphere of Mars is covered with craters. The northern hemisphere is smoother and appears to be covered by lava flows. Some scientists have proposed that the lava flows were caused by the impact of an object about 2,000 km in diameter.

 Key Concept Check Describe the atmosphere of each inner planet.

Figure 11 Gullies such as these might have been formed by the flow of liquid water. ▼

Lesson 2 Review

Assessment Online Quiz

Visual Summary

The terrestrial planets include Mercury, Venus, Earth, and Mars.

The inner planets all are made of rocks and minerals, but they have different characteristics. Earth is the only planet with liquid water.

The greenhouse effect greatly increases the surface temperature of Venus.

FOLDABLES

Use your lesson Foldable to review the lesson. Save your Foldable for the project at the end of the chapter.

What do you think NOW?

You first read the statements below at the beginning of the chapter.

3. Earth is the only inner planet that has a moon.
4. Venus is the hottest planet in the solar system.

Did you change your mind about whether you agree or disagree with the statements? Rewrite any false statements to make them true.

Use Vocabulary

1. **Define** *greenhouse effect* in your own words.

Understand Key Concepts

2. **Explain** why Venus is hotter than Mercury, even though Mercury is closer to the Sun.

3. **Infer** Why could rovers be used to explore Mars but not Venus?

4. Which of the inner planets has the greatest mass?
   A. Mercury   C. Earth
   B. Venus   D. Mars

5. **Relate** Describe the relationship between an inner planet's distance from the Sun and its period of revolution.

Interpret Graphics

6. **Infer** Which planet shown below is most likely able to support life now or was able to in the past? Explain your reasoning.

Mercury   Venus   Mars

7. **Compare and Contrast** Copy and fill in the table below to compare and contrast properties of Venus and Earth.

Planet	Similarities	Differences
Venus		
Earth		

Critical Thinking

8. **Imagine** How might the temperatures on Mercury be different if it had the same mass as Earth? Explain.

9. **Judge** Do you think the inner planets should be explored or should the money be spent on other things? Justify your opinion.

774 • Chapter 21 EVALUATE

Inquiry Skill Practice: Graphing Data

25 minutes

What can we learn about planets by graphing their characteristics?

Scientists collect and analyze data, and draw conclusions based on data. They are particularly interested in finding trends and relationships in data. One commonly used method of finding relationships is by graphing data. Graphing allows different types of data be to seen in relation to one another.

Learn It

Scientists know that some properties of the planets are related. **Graphing data** makes the relationships easy to identify. The graphs can show mathematical relationships such as direct and inverse relationships. Often, however, the graphs show that there is no relationship in the data.

Try It

1. You will plot two graphs that explore the relationships in data. The first graph compares a planet's distance from the Sun and its orbital period. The second graph compares a planet's distance from the Sun and its radius. Make a prediction about how these two sets of data are related, if at all. The data is shown in the table below.

Planet	Average Distance From the Sun (AU)	Orbital Period (yr)	Planet Radius (km)
Mercury	0.39	0.24	2440
Venus	0.72	0.62	6051
Earth	1.00	1.0	6378
Mars	1.52	1.9	3397
Jupiter	5.20	11.9	71,492
Saturn	9.58	29.4	60,268
Uranus	19.2	84.0	25,559
Neptune	30.1	164	24,764

2. Use the data in the table to plot a line graph showing orbital period versus average distance from the Sun. On the x-axis, plot the planet's distance from the Sun. On the y-axis, plot the planet's orbital period. Make sure the range of each axis is suitable for the data to be plotted, and clearly label each planet's data point.

3. Use the data in the table to plot a line graph showing planet radius versus average distance from the Sun. On the y-axis, plot the planet's radius. Make sure the range of each axis is suitable for the data to be plotted, and clearly label each planet's data point.

Apply It

4. Examine the *Orbital Period v. Distance from the Sun* graph. Does the graph show a relationship? If so, describe the relationship between a planet's distance from the Sun and its orbital period in your Science Journal.

5. Examine the *Planet Radius v. Distance from the Sun* graph. Does the graph show a relationship? If so, describe the relationship between a planet's distance from the Sun and its radius.

6. **Key Concept** Identify one or two characteristics the inner planets share that you learned from your graphs.

Lesson 3

Reading Guide

Key Concepts 🗝
ESSENTIAL QUESTIONS

- How are the outer planets similar?
- What are the outer planets made of?

Vocabulary
Galilean moons p. 779

 Multilingual eGlossary

The Outer Planets

Inquiry What's below?

Clouds often prevent airplane pilots from seeing the ground below. Similarly, clouds block the view of Jupiter's surface. What do you think is below Jupiter's colorful cloud layer? The answer might surprise you—Jupiter is not at all like Earth.

Inquiry Launch Lab

15 minutes

How do we see distant objects in the solar system?

Some of the outer planets were discovered hundreds of years ago. Why weren't all planets discovered?

1. Read and complete a lab safety form.
2. Use a **meterstick, masking tape,** and the **data table** to mark and label the position of each object on the tape on the floor along a straight line.
3. Shine a **flashlight** from "the Sun" horizontally along the tape.
4. Have a partner hold a page of this **book** in the flashlight beam at each planet location. Record your observations in your Science Journal.

Object	Distance from Sun (cm)
Sun	0
Jupiter	39
Saturn	71
Uranus	143
Neptune	295

Think About This

1. What happens to the image of the page as you move away from the flashlight?
2. **Key Concept** Why do you think it is more difficult to observe the outer planets than the inner planets?

The Gas Giants

Have you ever seen water drops on the outside of a glass of ice? They form because water vapor in the air changes to a liquid on the cold glass. Gases also change to liquids at high pressures. These properties of gases affect the outer planets.

The outer planets, shown in **Figure 12,** are called the gas giants because they are primarily made of hydrogen and helium. These elements are usually gases on Earth.

The outer planets have strong gravitational forces due to their large masses. The strong gravity creates tremendous atmospheric pressure that changes gases to liquids. Thus, the outer planets mainly have liquid interiors. In general, the outer planets have a thick gas and liquid layer covering a small, solid core.

Key Concept Check How are the outer planets similar?

Figure 12 The outer planets are primarily made of gases and liquids.

Visual Check Which outer planet is the largest?

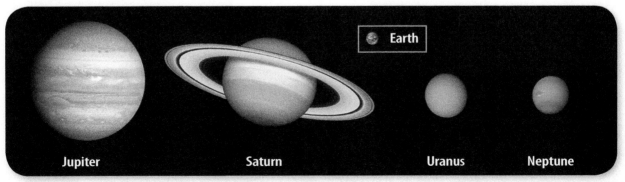

Lesson 3
777
EXPLORE

Jupiter

Figure 13 describes Jupiter, the largest planet in the solar system. Jupiter's diameter is more than 11 times larger than the diameter of Earth. Its mass is more than twice the mass of all the other planets combined. One way to understand just how big Jupiter is is to realize that more than 1,000 Earths would fit within this gaseous planet's volume.

Jupiter takes almost 12 Earth years to complete one orbit. Yet, it rotates faster than any other planet. Its period of rotation is less than 10 hours. Jupiter and all the outer planets have a ring system.

Jupiter's Atmosphere

The atmosphere on Jupiter is about 90 percent hydrogen and 10 percent helium and is about 1,000 km deep. Within the atmosphere are layers of dense, colorful clouds. Because Jupiter rotates so quickly, these clouds stretch into colorful, swirling bands. The Great Red Spot on the planet's surface is a storm of swirling gases.

Jupiter's Structure

Overall, Jupiter is about 80 percent hydrogen and 20 percent helium with small amounts of other materials. The planet is a ball of gas swirling around a thick liquid layer that conceals a solid core. About 1,000 km below the outer edge of the cloud layer, the pressure is so great that the hydrogen gas changes to liquid. This thick layer of liquid hydrogen surrounds Jupiter's core. Scientists do not know for sure what makes up the core. They suspect that the core is made of rock and iron. The core might be as large as Earth and could be 10 times more massive.

 Key Concept Check Describe what makes up each of Jupiter's three distinct layers.

FOLDABLES

Make a four-door book. Label each door with the name of an outer planet. Use the book to organize your notes on the outer planets.

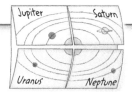

Figure 13 Jupiter is mainly hydrogen and helium. Throughout most of the planet, the pressure is high enough to change the hydrogen gas into a liquid.

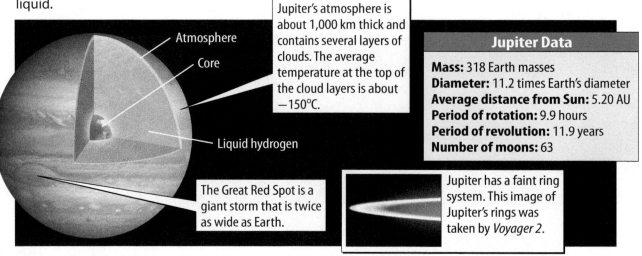

Jupiter's atmosphere is about 1,000 km thick and contains several layers of clouds. The average temperature at the top of the cloud layers is about −150°C.

Liquid hydrogen

The Great Red Spot is a giant storm that is twice as wide as Earth.

Jupiter Data

Mass: 318 Earth masses
Diameter: 11.2 times Earth's diameter
Average distance from Sun: 5.20 AU
Period of rotation: 9.9 hours
Period of revolution: 11.9 years
Number of moons: 63

Jupiter has a faint ring system. This image of Jupiter's rings was taken by *Voyager 2*.

The Moons of Jupiter

Jupiter has at least 63 moons, more than any other planet. Jupiter's four largest moons were first discovered by Galileo Galilei in 1610. *The four largest moons of Jupiter—Io, Europa, Ganymede, and Callisto—are known as the* **Galilean moons.** The Galilean moons all are made of rock and ice. The moons Ganymede, Callisto, and Io are larger than Earth's Moon. Collisions between Jupiter's moons and meteorites likely resulted in the particles that make up the planet's faint rings.

Saturn

Saturn is the sixth planet from the Sun. Like Jupiter, Saturn rotates rapidly and has horizontal bands of clouds. Saturn is about 90 percent hydrogen and 10 percent helium. It is the least dense planet. Its density is less than that of water.

Saturn's Structure

Saturn is made mostly of hydrogen and helium with small amounts of other materials. As shown in **Figure 14,** Saturn's structure is similar to Jupiter's structure—an outer gas layer, a thick layer of liquid hydrogen, and a solid core.

The ring system around Saturn is the largest and most complex in the solar system. Saturn has seven bands of rings, each containing thousands of narrower ringlets. The main ring system is over 70,000 km wide, but it is likely less than 30 m thick. The ice particles in the rings are possibly from a moon that was shattered in a collision with another icy object.

 Key Concept Check Describe what makes up Saturn and its ring system.

Math Skills

Ratios

A ratio is a quotient—it is one quantity divided by another. Ratios can be used to compare distances. For example, Jupiter is 5.20 AU from the Sun, and Neptune is 30.05 from the Sun. Divide the larger distance by the smaller distance:

$$\frac{30.05 \text{ AU}}{5.20 \text{ AU}} = 5.78$$

Neptune is 5.78 times farther from the Sun than Jupiter.

Practice

How many times farther from the Sun is Uranus (distance = 19.20 AU) than Saturn (distance = 9.58 AU)?

 Review

- **Math Practice**
- **Personal Tutor**

Figure 14 Like Jupiter, Saturn is mainly hydrogen and helium. Saturn's rings are one of the most noticeable features of the solar system.

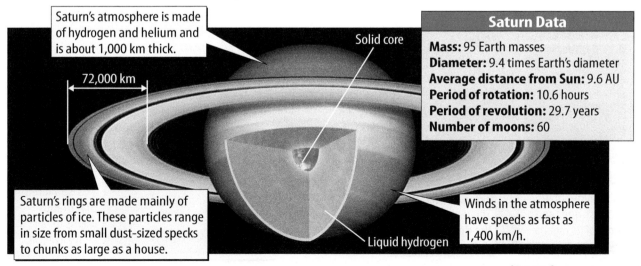

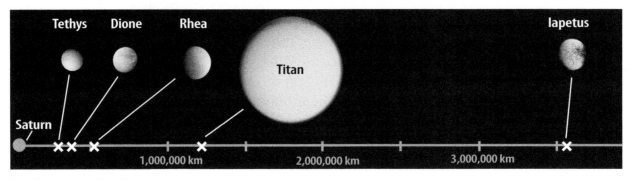

▲ Figure 15 The five largest moons of Saturn are shown above drawn to scale. Titan is Saturn's largest moon.

Saturn's Moons

Saturn has at least 60 moons. The five largest moons, Titan, Rhea, Dione, Iapetus, and Tethys, are shown in **Figure 15.** Most of Saturn's moons are chunks of ice less than 10 km in diameter. However, Titan is larger than the planet Mercury. Titan is the only moon in the solar system with a dense atmosphere. In 2005, the *Cassini* orbiter released the *Huygens* (HOY guns) probe that landed on Titan's surface.

WORD ORIGIN
probe
from Medieval Latin *proba*, means "examination"

Uranus

Uranus, shown in **Figure 16,** is the seventh planet from the Sun. It has a system of narrow, dark rings and a diameter about four times that of Earth. *Voyager 2* is the only space probe to explore Uranus. The probe flew by the planet in 1986.

Uranus has a deep atmosphere composed mostly of hydrogen and helium. The atmosphere also contains a small amount of methane. Beneath the atmosphere is a thick, slushy layer of water, ammonia, and other materials. Uranus might also have a solid, rocky core.

 Key Concept Check Identify the substances that make up the atmosphere and the thick slushy layer on Uranus.

Figure 16 Uranus is mainly gas and liquid, with a small solid core. Methane gas in the atmosphere gives Uranus a bluish color. ▼

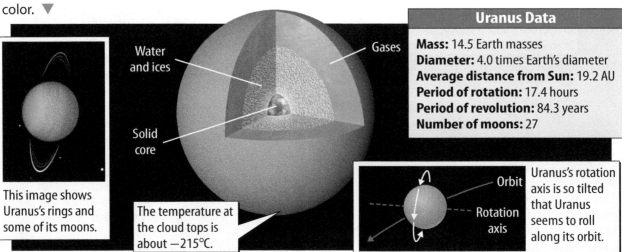

This image shows Uranus's rings and some of its moons.

The temperature at the cloud tops is about −215°C.

Uranus Data
Mass: 14.5 Earth masses
Diameter: 4.0 times Earth's diameter
Average distance from Sun: 19.2 AU
Period of rotation: 17.4 hours
Period of revolution: 84.3 years
Number of moons: 27

Uranus's rotation axis is so tilted that Uranus seems to roll along its orbit.

Uranus's Axis and Moons

Figure 16 shows that Uranus has a tilted axis of rotation. In fact, it is so tilted that the planet moves around the Sun like a rolling ball. This sideways tilt might have been caused by a collision with an Earth-sized object.

Uranus has at least 27 moons. The two largest moons, Titania and Oberon, are considerably smaller than Earth's moon. Titania has an icy cracked surface that once might have been covered by an ocean.

Neptune

Neptune, shown in **Figure 17,** was discovered in 1846. Like Uranus, Neptune's atmosphere is mostly hydrogen and helium, with a trace of methane. Its interior also is similar to the interior of Uranus. Neptune's interior is partially frozen water and ammonia with a rock and iron core.

Neptune has at least 13 moons and a faint, dark ring system. Its largest moon, Triton, is made of rock with an icy outer layer. It has a surface of frozen nitrogen and geysers that erupt nitrogen gas.

 Key Concept Check How does the atmosphere and interior of Neptune compare with that of Uranus?

How do Saturn's moons affect its rings?

In this lab, sugar models Saturn's rings. How might Saturn's moons affect its rings?

1. Read and complete a lab safety form.
2. Hold two **sharpened pencils** with their points even and then **tape** them together.
3. Insert a third pencil into the hole in a **record.** Hold the pencil so the record is in a horizontal position.
4. Have your partner sprinkle **sugar** evenly over the surface of the record. Hold the taped pencils vertically over the record so that the tips rest in the record's grooves.
5. Slowly turn the record. In your Science Journal, record what happens to the sugar.

Analyze and Conclude

1. **Compare and Contrast** What feature of Saturn's rings do the pencils model?
2. **Infer** What do you think causes the spaces between the rings of Saturn?
3. **Key Concept** What would have to be true for a moon to interact in this way with Saturn's rings?

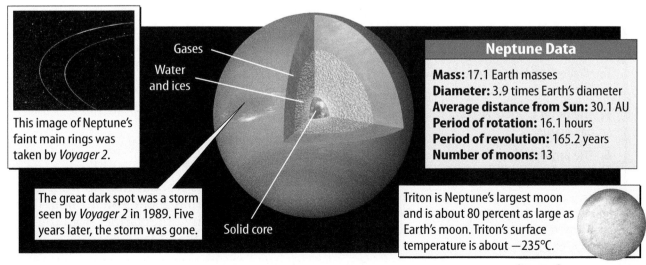

Figure 17 The atmosphere of Neptune is similar to that of Uranus—mainly hydrogen and helium with a trace of methane. The dark circular areas on Neptune are swirling storms. Winds on Neptune sometimes exceed 1,000 km/h.

This image of Neptune's faint main rings was taken by *Voyager 2*.

The great dark spot was a storm seen by *Voyager 2* in 1989. Five years later, the storm was gone.

Neptune Data

Mass: 17.1 Earth masses
Diameter: 3.9 times Earth's diameter
Average distance from Sun: 30.1 AU
Period of rotation: 16.1 hours
Period of revolution: 165.2 years
Number of moons: 13

Triton is Neptune's largest moon and is about 80 percent as large as Earth's moon. Triton's surface temperature is about −235°C.

Lesson 3 Review

Assessment | Online Quiz

Visual Summary

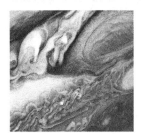

All of the outer planets are primarily made of materials that are gases on Earth. Colorful clouds of gas cover Saturn and Jupiter.

Jupiter is the largest outer planet. Its four largest moons are known as the Galilean moons.

Uranus has an unusual tilt, possibly due to a collision with a large object.

FOLDABLES

Use your lesson Foldable to review the lesson. Save your Foldable for the project at the end of the chapter.

What do you think NOW?

You first read the statements below at the beginning of the chapter.

5. The outer planets also are called the gas giants.
6. The atmospheres of Saturn and Jupiter are mainly water vapor.

Did you change your mind about whether you agree or disagree with the statements? Rewrite any false statements to make them true.

Use Vocabulary

1. **Identify** What are the four Galilean moons of Jupiter?

Understand Key Concepts

2. **Contrast** How are the rings of Saturn different from the rings of Jupiter?

3. Which planet's rings probably formed from a collision between an icy moon and another icy object?
   A. Jupiter   C. Saturn
   B. Neptune   D. Uranus

4. **List** the outer planets by increasing mass.

Interpret Graphics

5. **Infer** from the diagram below how Uranus's tilted axis affects its seasons.

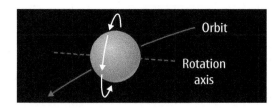

6. **Organize Information** Copy the organizer below and use it to list the outer planets.

Critical Thinking

7. **Predict** what would happen to Jupiter's atmosphere if its gravitational force suddenly decreased. Explain.

8. **Evaluate** Is life more likely on a dry and rocky moon or on an icy moon? Explain.

Math Skills

Math Practice

9. **Calculate** Mars is about 1.52 AU from the Sun, and Saturn is about 9.58 AU from the Sun. How many times farther from the Sun is Saturn than Mars?

Pluto

CAREERS in SCIENCE

What in the world is it?

Since Pluto's discovery in 1930, students have learned that the solar system has nine planets. But in 2006, the number of planets was changed to eight. What happened?

Neil deGrasse Tyson is an astrophysicist at the American Museum of Natural History in New York City. He and his fellow Museum scientists were among the first to question Pluto's classification as a planet. One reason was that Pluto is smaller than six moons in our solar system, including Earth's moon. Another reason was that Pluto's orbit is more oval-shaped, or elliptical, than the orbits of other planets. Also, Pluto has the most tilted orbit of all planets—17 degrees out of the plane of the solar system. Finally, unlike other planets, Pluto is mostly ice.

Tyson also questioned the definition of a planet—an object that orbits the Sun. Then shouldn't comets be planets? In addition, he noted that when Ceres, an object orbiting the Sun between Jupiter and Mars, was discovered in 1801, it was classified as a planet. But, as astronomers discovered more objects like Ceres, it was reclassified as an asteroid. Then, during the 1990s, many space objects similar to Pluto were discovered. They orbit the Sun beyond Neptune's orbit in a region called the Kuiper belt.

These new discoveries led Tyson and others to conclude that Pluto should be reclassified. In 2006, the International Astronomical Union agreed. Pluto was reclassified as a dwarf planet—an object that is spherical in shape and orbits the Sun in a zone with other objects. Pluto lost its rank as smallest planet, but became "king of the Kuiper belt."

Pluto TIME LINE

1930
Astronomer Clyde Tombaugh discovers a ninth planet, Pluto.

1992
The first object is discovered in the Kuiper belt.

July 2005
Eris—a Pluto-sized object—is discovered in the Kuiper belt.

January 2006
NASA launches *New Horizons* spacecraft, expected to reach Pluto in 2015.

August 2006
Pluto is reclassified as a dwarf planet.

▶ Neil deGrasse Tyson is director of the Hayden Planetarium at the American Museum of Natural History.

This illustration shows what Pluto might look like if you were standing on one of its moons.

It's Your Turn

RESEARCH With a group, identify the different types of objects in our solar system. Consider size, composition, location, and whether the objects have moons. Propose at least two different ways to group the objects.

Lesson 3 EXTEND

Lesson 4

Reading Guide

Key Concepts
ESSENTIAL QUESTIONS

- What is a dwarf planet?
- What are the characteristics of comets and asteroids?
- How does an impact crater form?

Vocabulary

meteoroid p. 788
meteor p. 788
meteorite p. 788
impact crater p. 788

 Multilingual eGlossary

Dwarf Planets and Other Objects

 Will it return?

You would probably remember a sight like this. This image of comet C/2006 P1 was taken in 2007. The comet is no longer visible from Earth. Believe it or not, many comets appear then reappear hundreds to millions of years later.

Inquiry Launch Lab

15 minutes

How might asteroids and moons form?

In this activity, you will explore one way moons and asteroids might have formed.

1. Read and complete a lab safety form.
2. Form a small ball from **modeling clay** and roll it in **sand.**
3. Press a thin layer of modeling clay around a **marble.**
4. Tie equal lengths of **string** to each ball. Hold the strings so the balls are above a **sheet of paper.**
5. Have someone pull back the marble so that its string is parallel to the tabletop and then release it. Record the results in your Science Journal.

Think About This

1. If the collision you modeled occurred in space, what would happen to the sand?
2. **Key Concept** Infer one way scientists propose moons and asteroids formed.

Dwarf Planets

Ceres was discovered in 1801 and was called a planet until similar objects were discovered near it. Then it was called an asteroid. For decades after Pluto's discovery in 1930, it was called a planet. Then, similar objects were discovered, and Pluto lost its planet classification. What type of object is Pluto?

Pluto once was classified as a planet, but it is now classified as a dwarf planet. In 2006, the International Astronomical Union (IAU) adopted "dwarf planet" as a new category. The IAU defines a dwarf planet as an object that orbits the Sun, has enough mass and gravity to form a sphere, and has objects similar in mass orbiting near it or crossing its orbital path. Astronomers classify Pluto, Ceres, Eris, MakeMake (MAH kay MAH kay), and Haumea (how MAY uh) as dwarf planets. **Figure 18** shows four dwarf planets.

Key Concept Check Describe the characteristics of a dwarf planet.

Figure 18 Four dwarf planets are shown to scale. All dwarf planets are smaller than the Moon.

Dwarf Planets

Earth's Moon | Eris | Pluto | Makemake | Ceres

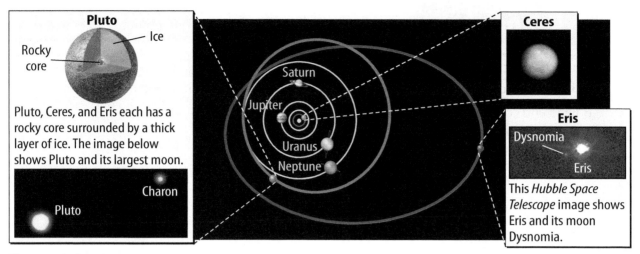

Figure 19 Because most dwarf planets are so far from Earth, astronomers do not have detailed images of them.

Visual Check Which dwarf planet orbits closest to Earth?

Make a layered book from two sheets of paper. Label it as shown. Use it to organize your notes on other objects in the solar system.

Ceres

Ceres, shown in **Figure 19,** orbits the Sun in the asteroid belt. With a diameter of about 950 km, Ceres is about one-fourth the size of the Moon. It is the smallest dwarf planet. Ceres might have a rocky core surrounded by a layer of water ice and a thin, dusty crust.

Pluto

Pluto is about two-thirds the size of the Moon. Pluto is so far from the Sun that its period of revolution is about 248 years. Like Ceres, Pluto has a rocky core surrounded by ice. With an average surface temperature of about −230°C, Pluto is so cold that it is covered with frozen nitrogen.

Pluto has three known moons. The largest moon, Charon, has a diameter that is about half the diameter of Pluto. Pluto also has two smaller moons, Hydra and Nix.

Eris

The largest dwarf planet, Eris, was discovered in 2003. Its orbit lasts about 557 years. Currently, Eris is three times farther from the Sun than Pluto is. The structure of Eris is probably similar to Pluto. Dysnomia (dis NOH mee uh) is the only known moon of Eris.

Makemake and Haumea

In 2008, the IAU designated two new objects as dwarf planets: Makemake and Haumea. Though smaller than Pluto, Makemake is one of the largest objects in a region of the solar system called the Kuiper (KI puhr) belt. The Kuiper belt extends from about the orbit of Neptune to about 50 AU from the Sun. Haumea is also in the Kuiper belt and is smaller than Pluto.

Reading Check Which dwarf planet is the largest? Which dwarf planet is the smallest?

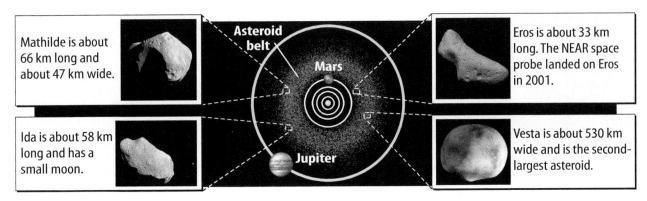

Figure 20 The asteroids that orbit the Sun in the asteroid belt are many sizes and shapes.

Asteroids

Recall from Lesson 1 that asteroids are pieces of rock and ice. Most asteroids orbit the Sun in the asteroid belt. The asteroid belt is between the orbits of Mars and Jupiter, as shown in **Figure 20**. Hundreds of thousands of asteroids have been discovered. The largest asteroid, Pallas, is over 500 km in diameter.

Asteroids are chunks of rock and ice that never clumped together like the rocks and ice that formed the inner planets. Some astronomers suggest that the strength of Jupiter's gravitational field might have caused the chunks to collide so violently, and they broke apart instead of sticking together. This means that asteroids are objects left over from the formation of the solar system.

 Key Concept Check Where do the orbits of most asteroids occur?

Comets

Recall that comets are mixtures of rock, ice, and dust. The particles in a comet are loosely held together by the gravitational attractions among the particles. As shown in **Figure 21,** comets orbit the Sun in long elliptical orbits.

The Structure of Comets

The solid, inner part of a comet is its nucleus, as shown in **Figure 21.** As a comet moves closer to the Sun, it absorbs thermal energy and can develop a bright tail. Heating changes the ice in the comet into a gas. Energy from the Sun pushes some of the gas and dust away from the nucleus and makes it glow. This produces the comet's bright tail and glowing nucleus, called a coma.

 Key Concept Check Describe the characteristics of a comet.

Figure 21 When energy from the Sun strikes the gas and dust in the comet's nucleus, it can create a two-part tail. The gas tail always points away from the Sun.

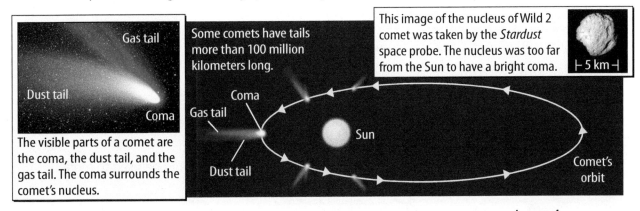

Short-Period and Long-Period Comets

A short-period comet takes less than 200 Earth years to orbit the Sun. Most short-period comets come from the Kuiper belt. A long-period comet takes more than 200 Earth years to orbit the Sun. Long-period comets come from a area at the outer edge of the solar system, called the Oort cloud. It surrounds the solar system and extends about 100,000 AU from the Sun. Some long-period comets take millions of years to orbit the Sun.

Meteoroids

Every day, many millions of particles called meteoroids enter Earth's atmosphere. *A* **meteoroid** *is a small, rocky particle that moves through space.* Most meteoroids are only about as big as a grain of sand. As a meteoroid passes through Earth's atmosphere, friction makes the meteoroid and the air around it hot enough to glow. *A* **meteor** *is a streak of light in Earth's atmosphere made by a glowing meteoroid.* Most meteoroids burn up in the atmosphere. However, some meteoroids are large enough that they reach Earth's surface before they burn up completely. When this happens, it is called a meteorite. *A* **meteorite** *is a meteoroid that strikes a planet or a moon.*

When a large meteoroite strikes a moon or planet, it often forms a bowl-shaped depression such as the one shown in **Figure 22.** *An* **impact crater** *is a round depression formed on the surface of a planet, moon, or other space object by the impact of a meteorite.* There are more than 170 impact craters on Earth.

Key Concept Check What causes an impact crater to form?

Figure 22 When a large meteorite strikes, it can form a giant impact crater like this 1.2-km wide crater in Arizona.

WORD ORIGIN
meteor
from Greek *meteoros*, means "high up"

Inquiry MiniLab
20 minutes

How do impact craters form?
In this lab, you will model the formation of an impact crater.

1. Pour a layer of **flour** about 3 cm deep in a **cake pan.**
2. Pour a layer of **cornmeal** about 1 cm deep on top of the flour.
3. One at a time, drop different-sized **marbles** into the mixture from the same height—about 15 cm. Record your observations in your Science Journal.

Analyze and Conclude

1. **Describe** the mixture's surface after you dropped the marbles.
2. **Recognize Cause and Effect** Based on your results, explain why impact craters on moons and planets differ.
3. **Key Concept** Explain how the marbles used in the activity could be used to model meteoroids, meteors, and meteorites.

Lesson 4 Review

Assessment Online Quiz

Visual Summary

An asteroid, such as Ida, is a chunk of rock and ice that orbits the Sun.

Comets, which are mixture of rock, ice, and dust, orbit the Sun. A comet's tail is caused by its interaction with the Sun.

When a large meteorite strikes a planet or moon, it often makes an impact crater.

FOLDABLES

Use your lesson Foldable to review the lesson. Save your Foldable for the project at the end of the chapter.

What do you think NOW?

You first read the statements below at the beginning of the chapter.

7. Asteroids and comets are mainly rock and ice.
8. A meteoroid is a meteor that strikes Earth.

Did you change your mind about whether you agree or disagree with the statements? Rewrite any false statements to make them true.

Use Vocabulary

1. **Define** *impact crater* in your own words.
2. **Distinguish** between a meteorite and a meteoroid.
3. **Use the term** *meteor* in a complete sentence.

Understand Key Concepts

4. Which produces an impact crater?
   A. comet   C. meteorite
   B. meteor   D. planet

5. **Reason** Are you more likely to see a meteor or a meteoroid? Explain.

6. **Differentiate** between objects located in the asteroid belt and objects located in the Kuiper belt.

Interpret Graphics

7. **Explain** why some comets have a two-part tail during portions of their orbit.

8. **Organize Information** Copy the table below and list the major characteristics of a dwarf planet.

Object	Defining Characteristic
Dwarf Planet	

Critical Thinking

9. **Compose** a paragraph describing what early sky observers might have thought when they saw a comet.

10. **Evaluate** Do you agree with the decision to reclassify Pluto as a dwarf planet? Defend your opinion.

Inquiry Lab

40 minutes

Scaling down the Solar System

Materials

2.25 in–wide register tape (several rolls)

meterstick

masking tape

colored markers

Safety

A scale model is a physical representation of something that is much smaller or much larger. Reduced-size scale models are made of very large things, such as the solar system. The scale used must reduce the actual size to a size reasonable for the model.

Question
What scale can you use to represent the distances between solar system objects?

Procedure

1. First, decide how big your solar system will be. Use the data given in the table to figure out how far apart the Sun and Neptune would be if a scale of 1 meter = 1 AU is used. Would a solar system based on that scale fit in the space you have available?

2. With your group determine the scale that results in a model that fits the available space. Larger models are usually more accurate, so choose a scale that produces the largest model that fits in the available space.

3. Once you have decided on a scale, copy the table in your Science Journal. Replace the word *(Scale)* in the third column of the table with the unit you have chosen. Then fill in the scaled distance for each planet.

Planet	Distance from the Sun (AU)	Distance from the Sun (Scale)
Mercury	0.39	
Venus	0.72	
Earth	1.00	
Mars	1.52	
Jupiter	5.20	
Saturn	9.54	
Uranus	19.18	
Neptune	30.06	

4. On register tape, mark the positions of objects in the solar system based on your chosen scale. Use a length of register tape that is slightly longer than the scaled distance between the Sun and Neptune.

5. Tape the ends of the register tape to a table or the floor. Mark a dot at one end of the paper to represent the Sun. Measure along the tape from the center of the dot to the location of Mercury. Mark a dot at this position and label it *Mercury*. Repeat this process for the remaining planets.

Analyze and Conclude

6. **Critique** There are many objects in the solar system. These objects have different sizes, structures, and orbits. Examine your scale model of the solar system. How accurate is the model? How could the model be changed to be more accurate?

7.  **The Big Idea** Pluto is a dwarf planet located beyond Neptune. Based on the pattern of distance data for the planets shown in the table, approximately how far from the Sun would you expect to find Pluto? Explain you reasoning.

8. **Calculate** What length of register tape is needed if a scale of 30 cm = 1 AU is used for the solar system model?

Lab Tips

☑ A scale is the ratio between the actual size of something and a representation of it.

☑ The distances between the planets and the Sun are average distances because planetary orbits are not perfect circles.

Communicate Your Results

Compare your model with other groups in your class by taping them all side-by-side. Discuss any major differences in your models. Discuss the difficulties in making the scale models much smaller.

 **Extension**

How can you build a scale model of the solar system that accurately shows both planetary diameters and distances? Describe how you would go about figuring this out.

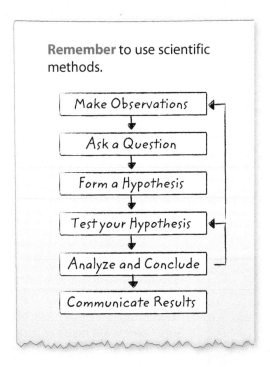

Chapter 21 Study Guide

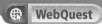

THE BIG IDEA: The solar system contains planets, dwarf planets, comets, asteroids, and other small solar system bodies.

Key Concepts Summary

Lesson 1: The Structure of the Solar System

- The inner planets are made mainly of solid materials. The outer planets, which are larger than the inner planets, have thick gas and liquid layers covering a small solid core.
- Astronomers measure vast distances in space in **astronomical units;** an astronomical unit is about 150 million km.
- The speed of each planet changes as it moves along its elliptical orbit around the Sun.

Lesson 2: The Inner Planets

- The inner planets—Mercury, Venus, Earth, and Mars—are made of rock and metallic materials.
- The **greenhouse effect** makes Venus the hottest planet.
- Mercury has no atmosphere. The atmospheres of Venus and Mars are almost entirely carbon dioxide. Earth's atmosphere is a mixture of gases and a small amount of water vapor.

Lesson 3: The Outer Planets

- The outer planets—Jupiter, Saturn, Uranus, and Neptune—are primarily made of hydrogen and helium.
- Jupiter and Saturn have thick cloud layers, but are mainly liquid hydrogen. Saturn's rings are largely particles of ice. Uranus and Neptune have thick atmospheres of hydrogen and helium.

Lesson 4: Dwarf Planets and Other Objects

- A dwarf planet is an object that orbits a star, has enough mass to pull itself into a spherical shape, and has objects similar in mass orbiting near it.
- An asteroid is a small rocky object that orbits the Sun. Comets are made of rock, ice, and dust and orbit the Sun in highly elliptical paths.
- The impact of a **meteorite** forms an **impact crater.**

Vocabulary

Lesson 1:
- **asteroid** p. 763
- **comet** p. 763
- **astronomical unit** p. 764
- **period of revolution** p. 764
- **period of rotation** p. 764

Lesson 2:
- **terrestrial planet** p. 769
- **greenhouse effect** p. 771

Lesson 3:
- **Galilean moons** p. 779

Lesson 4:
- **meteoroid** p. 788
- **meteor** p. 788
- **meteorite** p. 788
- **impact crater** p. 788

Study Guide

Review
- Personal Tutor
- Vocabulary eGames
- Vocabulary eFlashcards

FOLDABLES Chapter Project

Assemble your lesson Foldables as shown to make a Chapter Project. Use the project to review what you have learned in this chapter.

Use Vocabulary

Match each phrase with the correct vocabulary term from the Study Guide.

1. the time it takes an object to complete one rotation on its axis
2. the average distance from Earth to the Sun
3. the time it takes an object to travel once around the Sun
4. an increase in temperature caused by energy trapped by a planet's atmosphere
5. an inner planet
6. the four largest moons of Jupiter
7. a streak of light in Earth's atmosphere made by a glowing meteoroid

Concepts in Motion Interactive Concept Map

Link Vocabulary and Key Concepts

Copy this concept map, and then use vocabulary terms to complete the concept map.

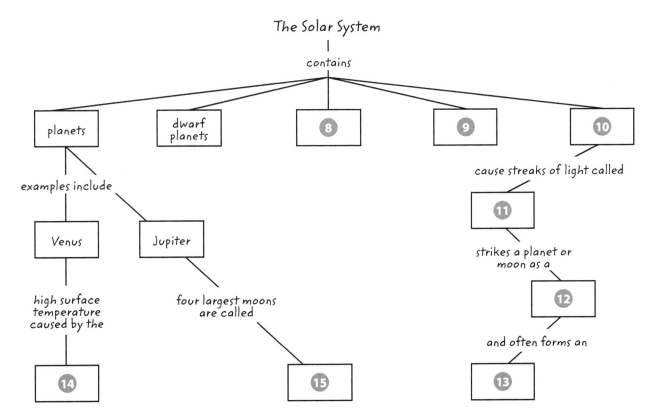

Chapter 21 Study Guide • 793

Chapter 21 Review

Understand Key Concepts

1. Which solar system object is the largest?
   A. Jupiter
   B. Neptune
   C. the Sun
   D. Saturn

2. Which best describes the asteroid belt?
   A. another name for the Oort cloud
   B. the region where comets originate
   C. large chunks of gas, dust, and ice
   D. millions of small rocky objects

3. Which describes a planet's speed as it orbits the Sun?
   A. It constantly decreases.
   B. It constantly increases.
   C. It does not change.
   D. It increases then decreases.

4. The diagram below shows a planet's orbit around the Sun. What does the blue arrow represent?
   A. the gravitational pull of the Sun
   B. the planet's orbital path
   C. the planet's path if Sun did not exist
   D. the planet's speed

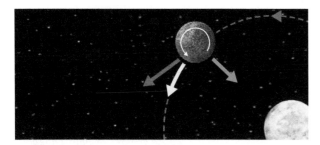

5. Which describes the greenhouse effect?
   A. effect of gravity on temperature
   B. energy emitted by the Sun
   C. energy trapped by atmosphere
   D. reflection of light from a planet

6. How are the terrestrial planets similar?
   A. similar densities
   B. similar diameters
   C. similar periods of rotation
   D. similar rocky surfaces

7. Which inner planet is the hottest?
   A. Earth
   B. Mars
   C. Mercury
   D. Venus

8. The photograph below shows how Earth appears from space. How does Earth differ from other inner planets?

   A. Its atmosphere contains large amounts of methane.
   B. Its period of revolution is much greater.
   C. Its surface is covered by large amounts of liquid water.
   D. Its surface temperature is higher.

9. Which two gases make up most of the outer planets?
   A. ammonia and helium
   B. ammonia and hydrogen
   C. hydrogen and helium
   D. methane and hydrogen

10. Which is true of the dwarf planets?
    A. more massive than nearby objects
    B. never have moons
    C. orbit near the Sun
    D. spherically shaped

11. Which is a bright streak of light in Earth's atmosphere?
    A. a comet
    B. a meteor
    C. a meteorite
    D. a meteoroid

12. Which best describes an asteroid?
    A. icy
    B. rocky
    C. round
    D. wet

Chapter Review

Critical Thinking

13 Relate changes in speed during a planet's orbit to the shape of the orbit and the gravitational pull of the Sun.

14 Compare In what ways are planets and dwarf planets similar?

15 Apply Like Venus, Earth's atmosphere contains carbon dioxide. What might happen on Earth if the amount of carbon dioxide in the atmosphere increases? Explain.

16 Defend A classmate states that life will someday be found on Mars. Defend the statement and offer a reason why life might exist on Mars.

17 Infer whether a planet with active volcanoes would have more or fewer craters than a planet without active volcanoes. Explain.

18 Support Use the diagram of the asteroid belt to support the explanation of how the belt formed.

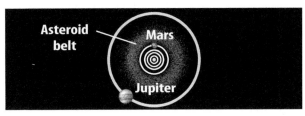

19 Evaluate The *Huygens* probe transmitted data about Titan for only 90 min. In your opinion, was this worth the effort of sending the probe?

20 Explain why Jupiter's moon Ganymede is not considered a dwarf planet, even though it is bigger than Mercury.

Writing in Science

21 Compose a pamphlet that describes how the International Astronomical Union classifies planets, dwarf planets, and small solar system objects.

REVIEW THE BIG IDEA

22 What kinds of objects are in the solar system? Summarize the types of space objects that make up the solar system and give at least one example of each.

23 The photo below shows part of Saturn's rings and two of its moons. Describe what Saturn and its rings are made of and explain why the other two objects are moons.

Math Skills

Use Ratios

Inner Planet Data

Planet	Diameter (% of Earth's diameter)	Mass (% of Earth's mass)	Average Distance from Sun (AU)
Mercury	38.3	5.5	0.39
Venus	95	81.5	0.72
Earth	100	100	1.00
Mars	53.2	10.7	1.52

24 Use the table above to calculate how many times farther from the Sun Mars is compared to Mercury.

25 Calculate how much greater Venus's mass is compared to Mercury's mass.

Chapter 21 Review • 795

Standardized Test Practice

Record your answers on the answer sheet provided by your teacher or on a sheet of paper.

Multiple Choice

1. Which is a terrestrial planet?
   A Ceres
   B Neptune
   C Pluto
   D Venus

2. An astronomical unit (AU) is the average distance
   A between Earth and the Moon.
   B from Earth to the Sun.
   C to the nearest star in the galaxy.
   D to the edge of the solar system.

3. Which is NOT a characteristic of ALL planets?
   A exceed the total mass of nearby objects
   B have a nearly spherical shape
   C have one or more moons
   D make an elliptical orbit around the Sun

Use the diagram below to answer question 4.

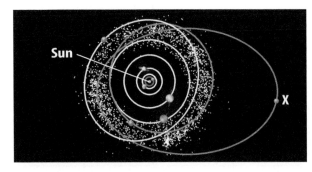

4. Which object in the solar system is marked by an X in the diagram?
   A asteroid
   B meteoroid
   C dwarf planet
   D outer planet

Use the diagram of Saturn below to answer questions 5 and 6.

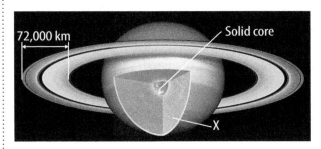

5. The thick inner layer marked X in the diagram above is made of which material?
   A carbon dioxide
   B gaseous helium
   C liquid hydrogen
   D molten rock

6. In the diagram, Saturn's rings are shown to be 72,000 km in width. Approximately how thick are Saturn's rings?
   A 30 m
   B 1,000 km
   C 14,000 km
   D 1 AU

7. Which are NOT found on Mercury's surface?
   A high cliffs
   B impact craters
   C lava flows
   D sand dunes

8. What is the primary cause of the extremely high temperatures on the surface of Venus?
   A heat rising from the mantle
   B lack of an atmosphere
   C proximity to the Sun
   D the greenhouse effect

Standardized Test Practice

Use the diagram below to answer question 9.

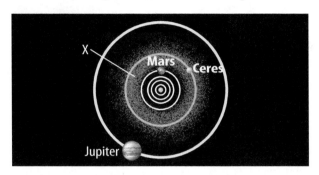

9 In the diagram above, which region of the solar system is marked by an *X*?
   A the asteroid belt
   B the dwarf planets
   C the Kuiper belt
   D the Oort cloud

10 What is a meteorite?
   A a surface depression formed by collision with a rock from space
   B a fragment of rock that strikes a planet or a moon
   C a mixture of ice, dust, and gas with a glowing tail
   D a small rocky particle that moves through space

11 What gives Mars its reddish color?
   A ice caps of frozen carbon dioxide
   B lava from Olympus Mons
   C liquid water in gullies
   D soil rich in iron oxide

Constructed Response

Use the table below to answer questions 12 and 13.

	Inner Planets	Outer Planets
Also called		
Relative size		
Main materials		
General structure		
Number of moons		

12 Copy the table and complete the first five rows to compare the features of the inner planets and outer planets.

13 In the blank row of the table, add another feature of the inner planets and outer planets. Then, describe the feature you have chosen.

14 What features of Earth make it suitable for supporting life as we know it?

15 How are planets, dwarf planets, and asteroids both similar and different?

NEED EXTRA HELP?														
If You Missed Question...	1	2	3	4	5	6	7	8	9	10	11	12	13	14	15
Go to Lesson...	2	1	1	1	3	3	2	2	1,4	4	2	2,3	2,3	2	1,4

Chapter 22

Stars and Galaxies

THE BIG IDEA What makes up the universe, and how does gravity affect the universe?

Inquiry What can't you see?

This photograph shows a small part of the universe. You can see many stars and galaxies in this image. But the universe also contains many things you cannot see.

- How do scientists study the universe?
- What makes up the universe?
- How does gravity affect the universe?

Get Ready to Read

What do you think?
Before you read, decide if you agree or disagree with each of these statements. As you read this chapter, see if you change your mind about any of the statements.

1. The night sky is divided into constellations.
2. A light-year is a measurement of time.
3. Stars shine because there are nuclear reactions in their cores.
4. Sunspots appear dark because they are cooler than nearby areas.
5. The more matter a star contains, the longer it is able to shine.
6. Gravity plays an important role in the formation of stars.
7. Most of the mass in the universe is in stars.
8. The Big Bang theory is an explanation of the beginning of the universe.

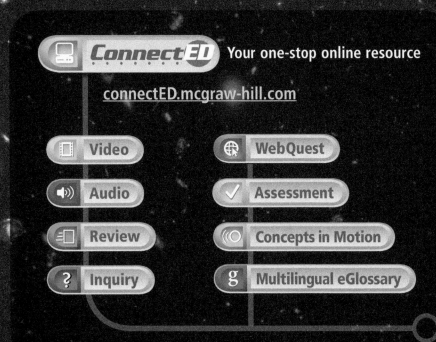

ConnectED Your one-stop online resource

connectED.mcgraw-hill.com

- Video
- WebQuest
- Audio
- Assessment
- Review
- Concepts in Motion
- Inquiry
- Multilingual eGlossary

Lesson 1

Reading Guide

Key Concepts 🔑
ESSENTIAL QUESTIONS

- How do astronomers divide the night sky?
- What can astronomers learn about stars from their light?
- How do scientists measure the distance and the brightness of objects in the sky?

Vocabulary
spectroscope p. 803
astronomical unit p. 804
light-year p. 804
apparent magnitude p. 805
luminosity p. 805

ⓖ **Multilingual eGlossary**

The View from Earth

Inquiry Where is this?

Unless you have visited a remote part of the country, you have probably never seen the sky look like this. It is similar to what the night sky looked like to your ancestors—before towns and cities brightened the sky.

Inquiry Launch Lab

20 minutes

How can you "see" invisible energy?

You see because of the Sun's light. You feel the heat of the Sun's energy. The Sun produces other kinds of energy that you can't directly see or feel.

1. Read and complete a lab safety form.
2. Put 5–6 **beads** into a **clear container.** Observe the color of the beads.
3. In a darkened room, shine light from a **flashlight** onto the beads for several seconds. Record your observations in your Science Journal. Repeat this step, exposing the beads to light from an **incandescent lightbulb** and a **fluorescent light.** Record your observations.
4. Stand outside in a shady spot for several seconds. Then expose the beads to direct sunlight. Record your observations.

Think About This

1. How did the light from the different light sources affect the color of the beads?
2. What do you think made the beads change color?
3. **Key Concept** How do you think invisible forms of light help scientists understand stars and other objects in the sky?

Looking at the Night Sky

Have you ever looked up at the sky on a clear, dark night and seen countless stars? If you have, you are lucky. Few people see a sky like that shown on the previous page. Lights from towns and cities make the night sky too bright for faint stars to be seen.

If you look at a clear night sky for a long time, the stars seem to move. But what you are really seeing is Earth's movement. Earth spins, or rotates, once every 24 hours. Day turns to night and then back to day as Earth rotates. Because Earth rotates from west to east, objects in the sky rise in the east and set in the west.

Earth spins on its axis, an imaginary line from the North Pole to the South Pole. The star Polaris is almost directly above the North Pole. As Earth spins, stars near Polaris appear to travel in a circle around Polaris, as shown in **Figure 1.** These stars never set when viewed from the northern hemisphere. They are always present in the night sky.

Figure 1 The stars around Polaris appear as streaks of light in this time-lapse photograph.

Lesson 1
EXPLORE

FOLDABLES

Make a horizontal two-tab book. Label it as shown. Use it to organize your notes on astronomy.

| How Scientists Divide the Night Sky | What Observations of the Stars Tell Scientists |

Naked-Eye Astronomy

You don't need expensive equipment to view the sky. *Naked-eye astronomy* means gazing at the sky with just your eyes, without binoculars or a telescope. Long before the telescope was invented, people observed stars to tell time and find directions. They learned about planets, seasons, and astronomical events merely by watching the sky. As you practice naked-eye astronomy, remember never to look directly at the Sun—it could damage your eyes.

Constellations

As people in ancient cultures gazed at the night sky, they saw patterns. The patterns resembled people, animals, or objects, such as the hunter and the dragon shown in **Figure 2.** The Greek astronomer Ptolemy (TAH luh mee) identified dozens of star patterns nearly 2,000 years ago. Today, these patterns and others like them are known as ancient constellations.

Present-day astronomers use many ancient constellations to divide the sky into 88 regions. Some of these regions, which are also called constellations, are shown in the sky map in **Figure 2.** Dividing the sky helps scientists communicate to others what area of sky they are studying.

 Key Concept Check How do astronomers divide the night sky?

Figure 2 Most modern constellations contain an ancient constellation.

Visual Check Why does east appear on the left and west appear on the right on the sky map?

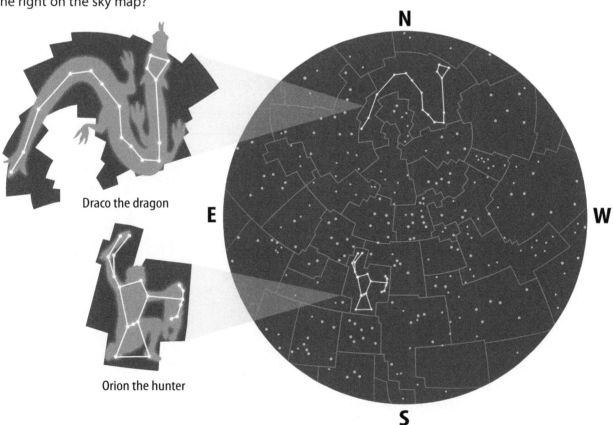

Draco the dragon

Orion the hunter

Electromagnetic Spectrum

Figure 3 Different parts of the electromagnetic spectrum have different wavelengths and different energies. You can see only a small part of the energy in these wavelengths.

Visual Check Which wavelength has the highest energy?

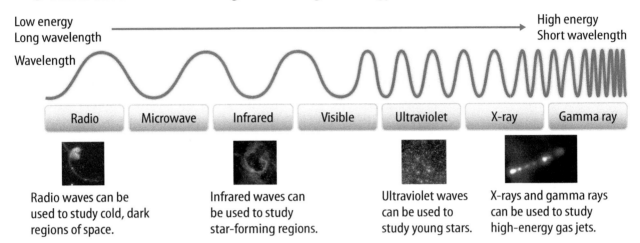

Radio waves can be used to study cold, dark regions of space.

Infrared waves can be used to study star-forming regions.

Ultraviolet waves can be used to study young stars.

X-rays and gamma rays can be used to study high-energy gas jets.

Telescopes

Telescopes can collect much more light than the human eye can detect. Visible light is just one part of the electromagnetic spectrum. As shown in **Figure 3**, the electromagnetic spectrum is a continuous range of wavelengths. Longer wavelengths have low energy. Shorter wavelengths have high energy. Different objects in space emit different ranges of wavelengths. The range of wavelengths that a star emits is the star's spectrum (plural, spectra).

Spectroscopes

Scientists study the spectra of stars using an instrument called a spectroscope. *A **spectroscope** spreads light into different wavelengths.* Using spectroscopes, astronomers can study stars' characteristics, including temperatures, compositions, and energies. For example, newly formed stars emit mostly radio and infrared waves, which have low energy. Exploding stars emit mostly high-energy ultraviolet waves and X-rays.

 Key Concept Check What can astronomers learn from a star's spectrum?

Inquiry MiniLab   20 minutes

How does light differ?

Light from the Sun is different from light from a lightbulb. How do the light sources differ?

1. Read and complete a lab safety form.
2. Follow instructions included with your **spectroscope**. Use it to observe various **light sources** around the classroom. Then use it to look at a bright part of the sky. ⚠ Do not look directly at the Sun.
3. Use **colored pencils** to draw what you see for each light source in your Science Journal.

Analyze and Conclude

1. **Compare and Contrast** What colors did you see for each light source? How did the colors differ?
2. **Key Concept** How might a spectroscope be used to learn about stars?

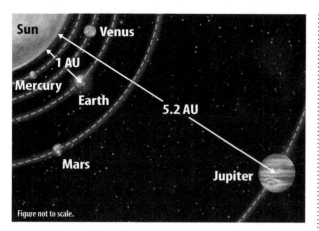

Figure 4 Measurements in the solar system are based on the average distance between Earth and the Sun—1 astronomical unit (AU). The most distant planet, Neptune, is 30 AU from the Sun.

WORD ORIGIN
parallax
from Greek *parallaxis*, means "alteration"

Math Skills

Use Proportions
Proportions can be used to calculate distances to astronomical objects. Light can travel nearly 10 trillion km in 1 year (y). How many years would it take light to reach Earth from a star that is 100 trillion km away?

1. Set up a proportion.

   $$\frac{10 \text{ trillion km}}{1 \text{ y}} = \frac{100 \text{ trillion km}}{x \text{ y}}$$

2. Cross multiply.

   10 trillion km × (*x*) y = 100 trillion km × 1 y

3. Solve for *x* by dividing both sides by 10 trillion km.

   $$x = \frac{100 \text{ trillion km}}{10 \text{ trillion km}} = 10 \text{ y}$$

Practice
How many years would it take light to reach Earth from a star 60 trillion km away?

- Review
- Math Practice
- Personal Tutor

Measuring Distances

Hold up your thumb at arm's length. Close one eye, and look at your thumb. Now open that eye, and close the other eye. Did your thumb seem to jump? This is an example of parallax. **Parallax** is the apparent change in an object's position caused by looking at it from two different points.

Astronomers use angles created by parallax to measure how far objects are from Earth. Instead of the eyes being the two points of view, they use two points in Earth's orbit around the Sun.

 Reading Check What is parallax?

Distances Within the Solar System

Because the universe is too large to be measured easily in meters or kilometers, astronomers use other units of measurement. For distances within the solar system, they use astronomical units (AU). *An* **astronomical unit** *is the average distance between Earth and the Sun, about 150 million km.* Astronomical units are convenient to use in the solar system because distances easily can be compared to the distance between Earth and the Sun, as shown in **Figure 4**.

Distances Beyond the Solar System

Astronomers measure distances to objects beyond the solar system using a larger distance unit—the light-year. Despite its name, a light-year measures distance, not time. *A* **light-year** *is the distance light travels in 1 year.* Light travels at a rate of about 300,000 km/s. That means 1 light-year is about 10 trillion km! Proxima Centauri, the nearest star to the Sun, is 4.2 light-years away.

Looking Back in Time

Because it takes time for light to travel, you see a star not as it is today, but as it was when light left it. At 4.2 light-years away, Proxima Centauri appears as it was 4.2 years ago. The farther away an object, the longer it takes for its light to reach Earth.

Measuring Brightness

When you look at stars, you can see that some are dim and some are bright. Astronomers measure the brightness of stars in two ways: by how bright they appear from Earth and by how bright they actually are.

Apparent Magnitude

Scientists measure how bright stars appear from Earth using a scale developed by the ancient Greek astronomer Hipparchus (hi PAR kus). Hipparchus assigned a number to every star he saw in the night sky, based on the star's brightness. Astronomers today call these numbers magnitudes. *The **apparent magnitude** of an object is a measure of how bright it appears from Earth.*

As shown in **Figure 5,** some objects have negative apparent magnitudes. That is because Hipparchus assigned a value of 1 to all of the brightest stars. He also did not assign values to the Sun, the Moon, or Venus. Astronomers later assigned negative numbers to the Sun, the Moon, Venus, and a few bright stars.

ACADEMIC VOCABULARY
apparent
(adjective) appearing to the eye or mind

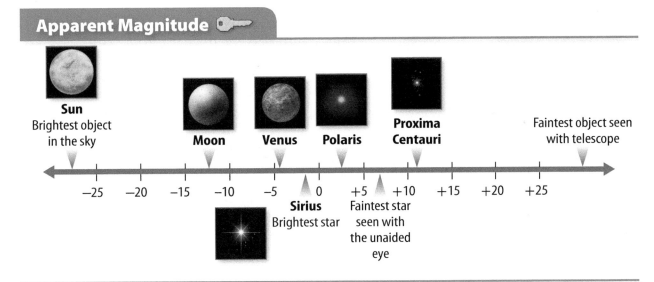

Absolute Magnitude

Stars can appear bright or dim depending on their distances from Earth. But stars also have actual, or absolute, magnitudes. **Luminosity** (lew muh NAH sih tee) *is the true brightness of an object.* The luminosity of a star, measured on an absolute magnitude scale, depends on the star's temperature and size, not its distance from Earth. A star's luminosity, apparent magnitude, and distance are related. If scientists know two of these factors, they can determine the third using mathematical formulas.

Key Concept Check How do scientists measure the brightness of stars?

Figure 5 The fainter a star or other object in the sky appears, the greater its apparent magnitude.

Visual Check What is the apparent magnitude of Sirius?

Lesson 1 Review

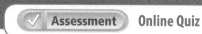

Visual Summary

Ancient people recognized patterns in the night sky. These patterns are known as the ancient constellations.

Different wavelengths of the electromagnetic spectrum carry different energies.

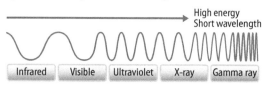

Astronomers measure distances within the solar system using astronomical units.

FOLDABLES

Use your lesson Foldable to review the lesson. Save your Foldable for the project at the end of the chapter.

What do you think NOW?

You first read the statements below at the beginning of the chapter.

1. The night sky is divided into constellations.
2. A light-year is a measurement of time.

Did you change your mind about whether you agree or disagree with the statements? Rewrite any false statements to make them true.

Use Vocabulary

1. A device that spreads light into different wavelengths is a(n) _____.

2. **Define** *astronomical unit* and *light-year* in your own words.

3. **Distinguish** between apparent magnitude and luminosity.

Understand Key Concepts

4. Which does a light-year measure?
   A. brightness   C. time
   B. distance   D. wavelength

5. **Describe** how scientists divide the sky.

Interpret Graphics

6. **Analyze** Which star in the diagram below appears the brightest from Earth?

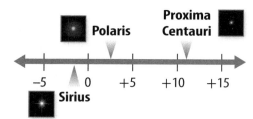

7. **Organize Information** Copy and fill in the graphic organizer below to list three things astronomers can learn from a star's light.

Critical Thinking

8. **Evaluate** why astronomers use modern constellation regions instead of ancient constellation patterns to divide the sky.

Math Skills

— Math Practice —

9. The Andromeda galaxy is about 25,000,000,000,000,000,000 km from Earth. How long does it take light to reach Earth from the Andromeda galaxy?

Inquiry Skill Practice
Interpret Scientific Illustrations — 30 minutes

How can you use scientific illustrations to locate constellations?

You might have heard that stars in the Big Dipper point to Polaris. The Big Dipper is a small star pattern in the larger constellation of Ursa Major. *Ursa Major* means "big bear" in Latin. It is the third-largest of the 88 modern constellations in the sky. Study the image of Ursa Major. Can you find the seven stars that form the Big Dipper? You can use a star finder to locate stars on any clear night of the year. The star finder also helps you see how constellations move across the sky.

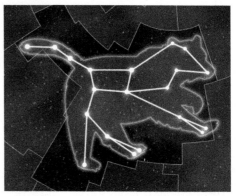

Materials

star chart

adhesive stars

graph paper

Learn It
Scientific illustrations can help you understand difficult or complicated subjects. **Interpret scientific illustrations** on the star finder to learn about the night sky.

Try It

1. Read and complete a lab safety form.

2. Read the user information provided with the star finder.

3. Rotate the wheel until the star finder is set to the day and time when you will be viewing the night sky. Observe how the ancient constellations marked on the star finder move.

4. Make a list of the bright stars, constellations, and planets you might be able to see in the sky.

5. Use the star finder outdoors on a clear night. As you hold the star finder overhead, be sure the arrows are pointing in the correct directions.

Apply It

6. What ancient constellations, planets, and stars were you able to see?

7. Did you locate Polaris? Why will you be able to see Polaris 6 months from now?

8. Which constellations won't you be able to see 6 months from now?

9. Why do stars appear to move?

10. How might ancient constellations have helped people in the past?

11. **Key Concept** How does dividing the sky into constellations help scientists study the sky?

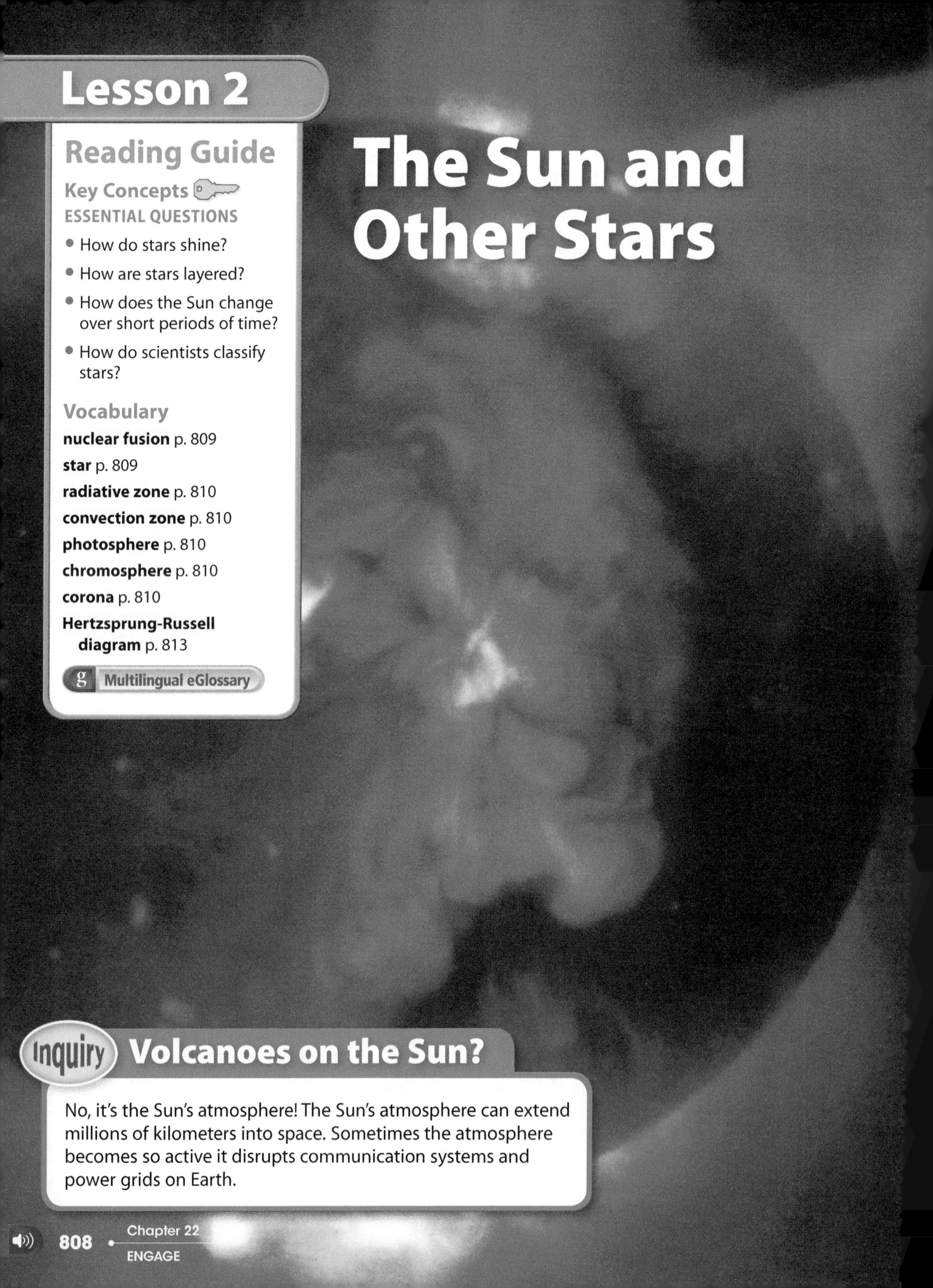

Lesson 2

Reading Guide

Key Concepts
ESSENTIAL QUESTIONS

- How do stars shine?
- How are stars layered?
- How does the Sun change over short periods of time?
- How do scientists classify stars?

Vocabulary

nuclear fusion p. 809
star p. 809
radiative zone p. 810
convection zone p. 810
photosphere p. 810
chromosphere p. 810
corona p. 810
Hertzsprung-Russell diagram p. 813

Multilingual eGlossary

The Sun and Other Stars

Inquiry Volcanoes on the Sun?

No, it's the Sun's atmosphere! The Sun's atmosphere can extend millions of kilometers into space. Sometimes the atmosphere becomes so active it disrupts communication systems and power grids on Earth.

Launch Lab

15 minutes

What are those spots on the Sun?

If you could see the Sun up close, what would it look like? Does it look the same all the time?

1. Examine a **collage of Sun images.** Notice the dates on which the pictures were taken.
2. Discuss with a partner what the dark spots might be and why they change position.
3. Select one spot. Estimate how long it took the spot to move completely across the surface of the Sun. Record your estimate in your Science Journal.

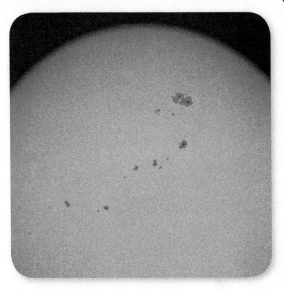

Think About This

1. What do you think the spots are?
2. Why do you think the spots move across the surface of the Sun?
3. **Key Concept** How do you think the Sun changes over days, months, and years?

How Stars Shine

The hotter something is, the more quickly its atoms move. As atoms move, they collide. If a gas is hot enough and its atoms move quickly enough, the nuclei of some of the atoms combine. **Nuclear fusion** *is a process that occurs when the nuclei of several atoms combine into one larger nucleus.*

Nuclear fusion releases a great amount of energy. This energy powers stars. *A* **star** *is a large ball of gas held together by gravity with a core so hot that nuclear fusion occurs.* A star's core can reach millions or hundreds of millions of degrees Celsius. When energy leaves a star's core, it travels throughout the star and radiates into space. As a result, the star shines.

 Key Concept Check How do stars shine?

Composition and Structure of Stars

The Sun is the closest star to Earth. Because it is so close, scientists can easily observe it. They can send probes to the Sun, and they can study its spectrum using spectroscopes on Earth-based telescopes. Spectra of the Sun and other stars provide information about *stellar* composition. The Sun and most stars are made almost entirely of hydrogen and helium gas. A star's composition changes slowly over time as hydrogen in its core fuses into more complex nuclei.

Make a vertical four-tab book. Label it as shown. Use it to organize your notes about the changing features of the Sun.

SCIENCE USE V. COMMON USE

stellar

Science Use anything related to stars

Common Use outstanding, exemplary

Layers of the Sun

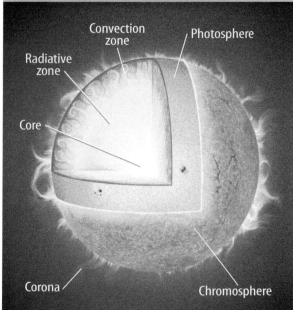

Figure 6 The Sun is divided into six layers.

Interior of Stars

When first formed, all stars fuse hydrogen into helium in their cores. Helium is denser than hydrogen, so it sinks to the inner part of the core after it forms.

The core is one of three interior layers of a typical star, as shown in the drawing of the Sun in **Figure 6.** *The* **radiative zone** *is a shell of cooler hydrogen above a star's core.* Hydrogen in this layer is dense. Light energy bounces from atom to atom as it gradually makes its way upward, out of the radiative zone.

Above the radiative zone is the **convection zone,** *where hot gas moves up toward the surface and cooler gas moves deeper into the interior.* Light energy moves quickly upward in the convection zone.

Key Concept Check What are the interior layers of a star?

Atmosphere of Stars

Beyond the convection zone are the three outer layers of a star. These layers make up a star's atmosphere. The **photosphere** *is the apparent surface of a star.* In the Sun, it is the dense, bright part you can see, where light energy radiates into space. From Earth, the Sun's photosphere looks smooth. But like the rest of the Sun, it is made of gas.

Above the photosphere are the two outer layers of a star's atmosphere. *The* **chromosphere** *is the orange-red layer above the photosphere,* as shown in **Figure 6.** *The* **corona** *is the wide, outermost layer of a star's atmosphere.* The temperature of the corona is higher than the photosphere or the chromosphere. It has an irregular shape and can extend outward for several million kilometers.

The Sun's Changing Features

The interior features of the Sun are stable over millions of years. But the Sun's atmosphere can change over years, months, or even minutes. Some of these features are illustrated in **Table 1** on the following page.

Inquiry MiniLab   20 minutes

Can you model the Sun's structure?

Making a two-dimensional model of the Sun can help you visualize its parts.

1. Read and complete a lab safety form.
2. Cut out each **Sun part.**
3. **Glue** the corona to a sheet of **black paper.** Glue the other pieces to the corona in this order: chromosphere, convection zone, radiative zone, core. Draw sunspots, solar flares, and prominences.

4. Glue only the top edge of the photosphere over the convection zone.

Analyze and Conclude

1. How would you model the Sun's rotation and convection? Add this to your model.
2. **Key Concept** How does this activity model a star's ability to shine?

Table 1 The Sun is dynamic. It changes over years, months, hours, and minutes.

Key Concept Check Which parts of the Sun change over short periods of time?

Table 1 Changing Features of the Sun

Sunspots Regions of strong magnetic activity are called sunspots. Cooler than the rest of the photosphere, sunspots appear as dark splotches on the Sun. They seem to move across the Sun as the Sun rotates. The number of sunspots changes over time. They follow a cycle, peaking in number every 11 years. An average sunspot is about the size of Earth.	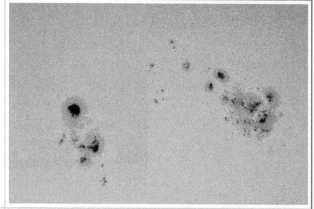
Prominences and Flares The loop shown here is a prominence. Prominences are clouds of gas that make loops and jets extending into the corona. They sometimes last for weeks. Flares are sudden increases in brightness often found near sunspots or prominences. They are violent eruptions that last from minutes to hours. Both prominences and flares begin at or just above the photosphere.	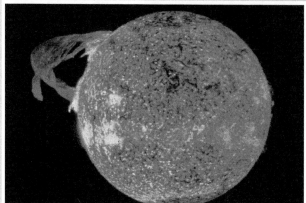
Coronal Mass Ejections (CMEs) Huge bubbles of gas ejected from the corona are coronal mass ejections (CMEs). They are much larger than flares and occur over the course of several hours. Material from a CME can reach Earth, occasionally causing a radio blackout or a malfunction in an orbiting satellite.	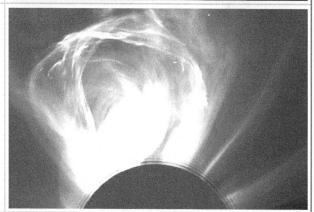
The Solar Wind Charged particles that stream continually away from the Sun create the solar wind. The solar wind passes Earth and extends to the edge of the solar system. Auroras are curtains of light created when particles from the solar wind or a CME interact with Earth's magnetic field. Auroras occur in both the northern and southern hemispheres. The northern lights are shown here.	

Figure 7 Open clusters (top) contain fewer than 1,000 stars. Globular clusters (bottom) can contain hundreds of thousands of stars.

WORD ORIGIN

globular
from Latin *globus*, means "round mass, sphere"

Groups of Stars

The Sun has no stellar companion. The star closest to the Sun is 4.2 light-years away. Many stars are single stars, such as the Sun. Most stars exist in multiple star systems bound by gravity.

The most common star system is a binary system, where two stars orbit each other. By studying the orbits of binary stars, astronomers can determine the stars' masses. Many stars exist in large groupings called clusters. Two types of star clusters—open clusters and globular clusters—are shown in Figure 7. Stars in a cluster all formed at about the same time and are the same distance from Earth. If astronomers determine the distance to or the age of one star in a cluster, they know the distance to or the age of every star in that cluster.

Classifying Stars

How do you classify a star? Which properties are important? Scientists classify stars according to their spectra. Recall that a star's spectrum is the light it emits spread out by wavelength. Stars have different spectra and different colors depending on their surface temperatures.

Temperature, Color, and Mass

Have you ever seen coals in a fire? Red coals are the coolest, and blue-white coals are the hottest. Stars are similar. Blue-white stars are hotter than red stars. Orange, yellow, and white stars are intermediate in temperature. Though there are exceptions, color in most stars is related to mass, as shown in Figure 8. Blue-white stars tend to have the most mass, followed by white stars, yellow stars, orange stars, and red stars.

 Reading Check How does star color relate to mass?

As shown in Figure 8, the Sun is tiny compared to large, blue-white stars. However, scientists suspect that most stars—as many as 90 percent—are smaller than the Sun. These stars are called red dwarfs. The smallest star in Figure 8 is a red dwarf.

Figure 8 The most massive stars are usually the hottest and are blue-white. The smallest stars tend to be cooler and red.

Sun

Hertzsprung-Russell Diagram

Figure 9 The H-R diagram plots luminosity against temperature. Most stars exist along the main sequence, the band that stretches from the upper left to the lower right.

Visual Check Where is the Sun on this diagram?

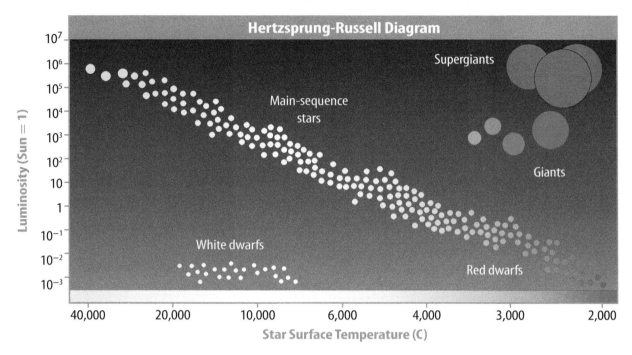

Hertzsprung-Russell Diagram

When scientists plot the temperatures of stars against their luminosities, the result is a graph like that shown in Figure 9. The **Hertzsprung-Russell diagram** (or H-R diagram) *is a graph that plots luminosity v. temperature of stars.* The *y*-axis of the H-R diagram displays increasing luminosity. The *x*-axis displays decreasing temperature.

The H-R diagram is named after two astronomers who developed it in the early 1900s. It is an important tool for categorizing stars. It also is an important tool for determining distances of some stars. If a star has the same temperature as a star on the H-R diagram, astronomers often can determine its luminosity. As you read earlier, if scientists know a star's luminosity, they can calculate its distance from Earth.

Key Concept Check What is the Hertzsprung-Russell diagram?

The Main Sequence

Most stars spend the majority of their lives on the main sequence. On the H-R diagram, main sequence stars form a curved line from the upper left corner to the lower right corner of the graph. The mass of a star determines both its temperature and its luminosity; the higher the mass the hotter and brighter the star. Because high-mass stars have more gravity pulling inward than low-mass stars, their cores have higher temperatures and produce and use more energy through fusion. High-mass stars have a shorter life span than low-mass stars. High-mass stars burn through their hydrogen much faster and move off the main sequence. A downside to a large-mass star is that the life span of the star is much shorter than average- or low-mass stars.

As shown in Figure 9, some groups of stars on the H-R diagram lie outside of the main sequence. These stars are no longer fusing hydrogen into helium in their cores. Some of these stars are cooler, but brighter and larger, such as supergiants. Other stars are dimmer and smaller, but much hotter, such as white dwarfs.

Lesson 2 Review

Assessment — Online Quiz
Inquiry — Virtual Lab

Visual Summary

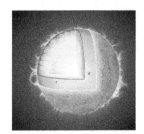
Hot gas moves up and cool gas moves down in the Sun's convection zone.

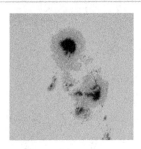
Sunspots are relatively dark areas on the Sun that have strong magnetic activity.

Globular clusters contain hundreds of thousands of stars.

Use your lesson Foldable to review the lesson. Save your Foldable for the project at the end of the chapter.

What do you think NOW?

You first read the statements below at the beginning of the chapter.

3. Stars shine because there are nuclear reactions in their cores.

4. Sunspots are dark because they are cooler than nearby areas.

Did you change your mind about whether you agree or disagree with the statements? Rewrite any false statements to make them true.

Use Vocabulary

1. The _____ is a graph that plots luminosity v. temperature.

2. **Use the term** *photosphere* in a sentence.

3. **Define** *star* in your own words.

Understand Key Concepts

4. Which part of a star extends millions of kilometers into space?
   A. chromosphere   C. photosphere
   B. corona   D. radiative zone

5. **Explain** how stars produce and release energy.

6. **Construct** an H-R diagram, and show the positions of the main sequence and the Sun.

Interpret Graphics

7. **Identify** Which star on the diagram below is hottest? Which is coolest? Which star represents the Sun?

8. **Organize Information** Copy and fill in the graphic organizer below to list the Sun's radiative zone, corona, convection zone, chromosphere, and photosphere in order outward from the core.

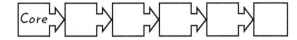

Critical Thinking

9. **Assess** why scientists monitor the Sun's changing features.

10. **Evaluate** In what way is the Sun an average star? In what way is it not an average star?

HOW IT WORKS

Viewing the Sun in 3-D

NASA's Solar Terrestrial Relations Observatory

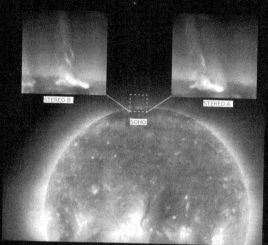

You might have used a telescope to look at objects far in the distance or to look at stars and planets. Although telescopes allow you to see a distant object in closer detail, you cannot see a three-dimensional view of objects in space. To get a three-dimensional view of the Sun, astronomers use two space telescopes. NASA's *Solar Terrestrial Relations Observatory* (STEREO) telescopes orbit the Sun in front of and behind Earth and give astronomers a 3-D view of the Sun. Why is this important?

If a coronal mass ejection (CME) erupts from the Sun, it can blast more than a billion tons of material into space. The powerful energy in a CME can damage satellites and power grids if Earth happens to be in its way. Before STEREO, scientists had only a straight-on view of CMEs approaching Earth. With STEREO, they have two different views. Each STEREO telescope carries several cameras that can detect many wavelengths. Scientists combine the pictures from each type of camera to make one 3-D image. In this way, they can track a CME from its emergence on the Sun all the way to its impact with Earth.

STEREO B is in orbit around the Sun behind Earth.

In January 2009, the telescopes were 90 degrees apart.

In February 2011, the craft will be 180 degrees apart.

Earth

STEREO A is in orbit around the Sun ahead of Earth.

It's Your Turn

RESEARCH AND REPORT How can power and satellite companies prepare for an approaching CME? Find out and write a short report on what you find. Share your findings with the class.

Lesson 2
EXTEND

Lesson 3

Reading Guide

Key Concepts
ESSENTIAL QUESTIONS
- How do stars form?
- How does a star's mass affect its evolution?
- How is star matter recycled in space?

Vocabulary
nebula p. 817
white dwarf p. 819
supernova p. 819
neutron star p. 820
black hole p. 820

g Multilingual eGlossary

Evolution of Stars

Inquiry Exploding Star?

No, this is a cloud of gas and dust where stars form. How do you think stars form? Do you think stars ever stop shining?

816 • Chapter 22
ENGAGE

Inquiry Launch Lab

20 minutes

Do stars have life cycles?
You might have learned about the life cycles of plants or animals. Do stars, such as the Sun, have life cycles? Before you find out, review the life cycle of a sunflower.

1. Read and complete a lab safety form.
2. Obtain an **envelope containing slips of paper** that explain the life cycle of a sunflower.
3. Use **colored pencils** to draw a sunflower in the middle of a piece of **paper,** or use a **glue stick** to glue a sunflower picture on the paper.
4. Using your knowledge of plant life cycles, arrange the slips of paper around the sunflower in the order in which the events listed on them occur. Draw arrows to show how the steps form a cycle.

Think About This

1. Does the life cycle of a sunflower have a beginning and an end? Explain your answer.
2. Do you think that every stage in the life cycle takes the same amount of time? Why or why not?
3. **Key Concept** How do you think the life cycle of a star compares to the life cycle of a sunflower? Do you think all stars have the same life cycle?

Life Cycle of a Star

Like living things, stars have life cycles. They are "born," and after millions or billions of years, they "die." Stars die in different ways, depending on their masses. But all stars—from white dwarfs to supergiants—form in the same way.

Nebulae and Protostars

Stars form deep inside clouds of gas and dust. *A cloud of gas and dust is a* **nebula** (plural, nebulae). Star-forming nebulae are cold, dense, and dark. Gravity causes the densest parts to collapse, forming regions called protostars. Protostars continue to contract, pulling in surrounding gas, until their cores are hot and dense enough for nuclear fusion to begin. As they contract, protostars produce enormous amounts of thermal energy.

Birth of a Star

Over many thousands of years, the energy produced by protostars heats the gas and dust surrounding them. Eventually, the surrounding gas and dust blows away, and the protostars become visible as stars. Some of this material might later become planets or other objects that orbit the star. During the star-formation process, nebulae glow brightly, as shown in the photograph on the previous page.

Key Concept Check How do stars form?

WORD ORIGIN
nebula
from Latin *nebula*, means "mist" or "little cloud"

FOLDABLES
Make a vertical five-tab book. Label it as shown. Use it to organize your notes on the life cycle of a star.

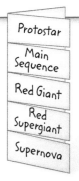

Lesson 3
EXPLORE

Main-Sequence Stars

Recall the main sequence of the Hertzsprung-Russell diagram. Stars spend most of their lives on the main sequence. A star becomes a main-sequence star as soon as it begins to fuse hydrogen into helium in the core. It remains on the main sequence for as long as it continues to fuse hydrogen into helium. Average-mass stars such as the Sun remain on the main sequence for billions of years. High-mass stars remain on the main sequence for only a few million years. Even though massive stars have more hydrogen than lower-mass stars, they process it at a much faster rate.

When a star's hydrogen supply is nearly gone, the star moves off the main sequence. It begins the next stage of its life cycle, as shown in **Figure 10**. Not all stars go through all phases in **Figure 10**. Lower-mass stars do not have enough mass to become supergiants.

Figure 10 Massive stars become red giants, then larger red giants, then red supergiants.

Visual Check Which element forms in only the most massive stars?

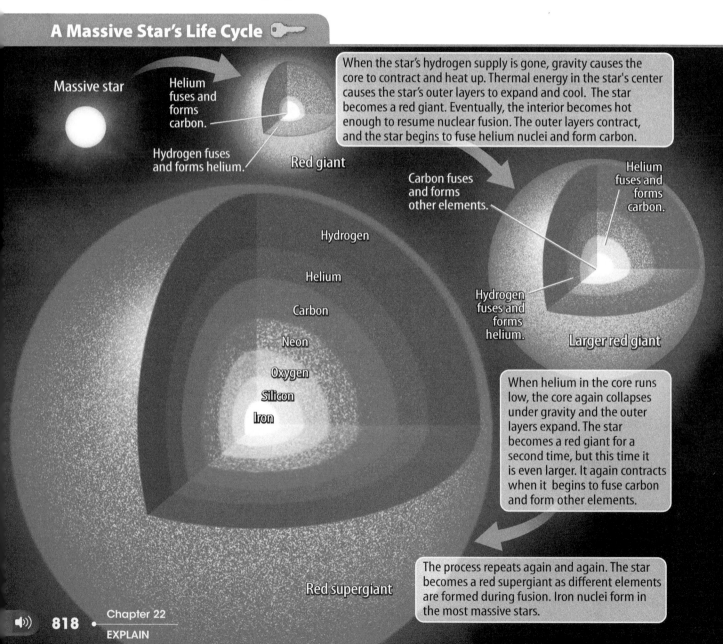

A Massive Star's Life Cycle

End of a Star

All stars form in the same way. But stars die in different ways, depending on their masses. Massive stars collapse and explode. Lower-mass stars die more slowly.

White Dwarfs

Average-mass stars, such as the Sun, do not have enough mass to fuse elements beyond helium. They do not get hot enough. After helium in their cores is gone, the stars cast off their gases, exposing their cores. The core becomes a **white dwarf**, *a hot, dense, slowly cooling sphere of carbon.*

What will happen to Earth and the solar system when the Sun runs out of fuel? When the Sun runs out of hydrogen, in about 5 billion years, it will become a red giant. Once helium fusion begins, the Sun will contract. When the helium is gone, the Sun will expand again, probably absorbing Mercury, Venus, and Earth and pushing Mars outward, as shown in **Figure 11**. Eventually, the Sun will become a white dwarf. Imagine the mass of the Sun squeezed a million times until it is the size of Earth. That's the size of a white dwarf. Scientists expect that all stars with masses less than 8–10 times that of the Sun will eventually become white dwarfs.

 Reading Check What will happen to Earth when the Sun runs out of fuel?

Supernovae

Stars with more than 10 times the mass of the Sun do not become white dwarfs. Instead, they explode. *A* **supernova** *(plural, supernovae) is an enormous explosion that destroys a star.* In the most massive stars, a supernova occurs when iron forms in the star's core. Iron is stable and does not fuse. After a star forms iron, it loses its internal energy source, and the core collapses quickly under the force of gravity. So much energy is released that the star explodes. When it explodes, a star can become one billion times brighter and form elements even heavier than iron.

Figure 11 In about 5 billion years, the Sun will become a red giant and then a white dwarf.

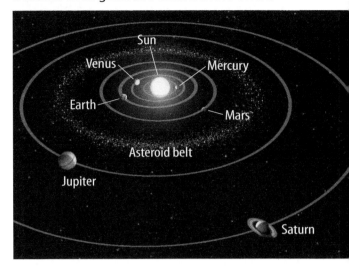

The Sun will remain on the main sequence for 5 billion more years.

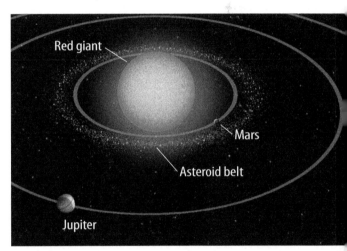

When the Sun becomes a red giant for the second time, the outer layers will probably absorb Earth and push Mars and Jupiter outward.

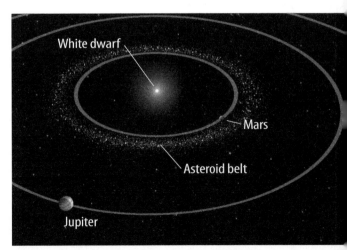

When the Sun becomes a white dwarf, the solar system will be a cold, dark place.

Lesson 3
EXPLAIN

REVIEW VOCABULARY
neutron
a neutral particle in the nucleus of an atom

Neutron Stars

Have you ever eaten cotton candy? A bag of cotton candy is made from just a few spoonfuls of spun sugar. Cotton candy is mostly air. Similarly, atoms are mostly empty space. During a supernova, the outer layers of the star are blown away and the core collapses under the heavy force of gravity. The space in atoms disappears as protons and electrons combine to form neutrons. *A **neutron star** is a dense core of neutrons that remains after a supernova.* Neutron stars are only about 20 km wide, with cores so dense that a teaspoonful would weigh more than 1 billion tons.

Black Holes

For the most massive stars, atomic forces holding neutrons together are not strong enough to overcome so much mass in such a small volume. Gravity is too strong, and the matter crushes into a black hole. *A **black hole** is an object whose gravity is so great that no light can escape.*

A black hole does not suck matter in like a vacuum cleaner. But its gravity is very strong because all of its mass is concentrated in a single point. Because astronomers cannot see a black hole, they only can infer its existence. For example, if they detect a star circling around something, but they cannot see what that something is, they infer it is a black hole.

 Key Concept Check How does a star's mass determine if it will become a white dwarf, a neutron star, or a black hole?

Inquiry MiniLab   15 minutes

How do astronomers detect black holes?

The only way astronomers can detect black holes is by studying the movement of objects nearby. How do black holes affect nearby objects?

1. Read and complete a lab safety form.
2. With a partner, make two stacks of **books** of equal height about 25 cm apart. Place a piece of **thin cardboard** on top of the books.
3. Spread some **staples** over the cardboard. Hold a **magnet** under the cardboard. Observe what happens to the staples.
4. While one student holds the magnet in place beneath the cardboard, the other student gently rolls a **small magnetic marble** across the cardboard. Repeat several times, rolling the marble in different pathways. Record your observations in your Science Journal.

Analyze and Conclude

1. **Infer** What did the pull of the magnet represent?
2. **Cause and Effect** How did the magnet affect the staples and the movement of the marble?
3. **Key Concept** How do black holes affect nearby objects?

Recycling Matter

At the end of a star's life cycle, much of its gas escapes into space. This gas is recycled. It becomes the building blocks of future generations of stars and planets.

Planetary Nebulae

You read that average-mass stars, such as the Sun, become white dwarfs. When a star becomes a white dwarf, it casts off hydrogen and helium gases in its outer layers, as shown in **Figure 12.** The expanding, cast-off matter of a white dwarf is a planetary nebula. Most of the star's carbon remains locked in the white dwarf. But the gases in the planetary nebula can be used to form new stars.

Planetary nebulae have nothing to do with planets. They are so named because early astronomers thought they were regions where planets were forming.

Supernova Remnants

During a supernova, a massive star comes apart. This sends a shock wave into space. The expanding cloud of dust and gas is called a supernova remnant. A supernova remnant is shown in **Figure 13.** Like a snowplow pushing snow in its path, a supernova remnant pushes on the gas and dust it encounters.

In a supernova, a star releases the elements that formed inside it during nuclear fusion. Almost all of the elements in the universe other than hydrogen and helium were created by nuclear reactions inside the cores of massive stars and released in supernovae. This includes the oxygen in air, the silicon in rocks, and the carbon in you.

 Key Concept Check How do stars recycle matter?

Gravity causes recycled gases and other matter to clump together in nebulae and form new stars and planets. As you will read in the next lesson, gravity also causes stars to clump together into even larger structures called galaxies.

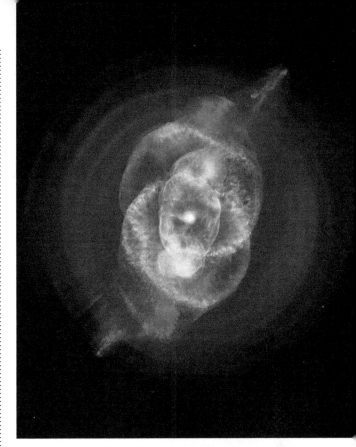

▲ **Figure 12** White dwarfs cast off helium and hydrogen as planetary nebulae. The gases can be used by new generations of stars.

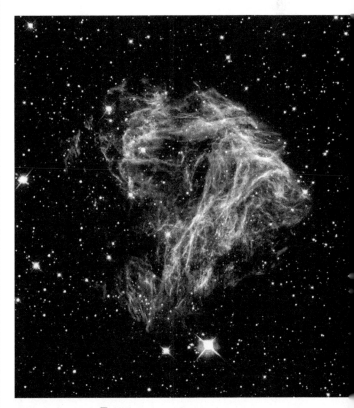

▲ **Figure 13** Many of the elements in you and in matter all around you were formed inside massive stars and released in supernovae.

Lesson 3 Review

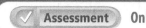

Visual Summary

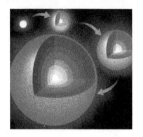

Iron is formed in the cores of the most massive stars.

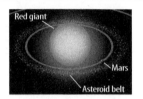

The Sun will become a red giant in about 5 billion years.

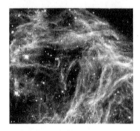

Matter is recycled in supernovae.

FOLDABLES

Use your lesson Foldable to review the lesson. Save your Foldable for the project at the end of the chapter.

What do you think NOW?

You first read the statements below at the beginning of the chapter.

5. The more matter a star contains, the longer it is able to shine.

6. Gravity plays an important role in the formation of stars.

Did you change your mind about whether you agree or disagree with the statements? Rewrite any false statements to make them true.

Use Vocabulary

1. Planetary nebulae are the expanding outer layers of a(n) _____.

2. **Define** *supernova* in your own words.

3. **Use the terms** *neutron star* and *black hole* in a sentence.

Understand Key Concepts

4. Which type of star will the Sun eventually become?
   A. neutron star   C. red supergiant
   B. red dwarf   D. white dwarf

5. **Explain** how supernovae recycle matter.

6. **Rank** black holes, neutron stars, and white dwarfs from smallest to largest. Then rank them from most massive to least massive.

Interpret Graphics

7. **Describe** details of the process occurring in the photo below.

8. **Organize Information** Copy and fill in the graphic organizer below to list what happens to a star following a supernova.

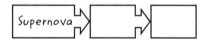

Critical Thinking

9. **Predict** whether the Sun will eventually become a black hole. Why or why not?

10. **Evaluate** why mass is so important in determining the evolution of a star.

Inquiry Skill Practice: Make and Use Graphs

45 minutes

How can graphing data help you understand stars?

How can you make sense of everything in the universe? Graphs help you organize information. The Hertzsprung-Russell diagram is a graph that plots the color, or temperature, of stars against their luminosities. What can you learn about stars by plotting them on a graph similar to the H-R diagram?

Materials

graph paper

Learn It

Displaying information on graphs makes it easier to see how objects are related. Lines on graphs show you patterns and enable you to make predictions. Graphs display a lot of information in an easily understandable form. In this activity, you will **make and use graphs,** plotting the temperature, the color, and the mass of stars.

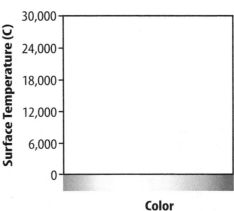

Try It

1. Using graph paper or your Science Journal, draw a graph like the one shown at right.

2. Use the color and temperature data in the table below to plot the position of each star on your graph. Mark the points by attaching adhesive stars to the graph.

3. If stars have similar data, plot them in a cluster. Label each star with its name.

4. Draw a curve that joins the data points as smoothly as possible.

5. Make another graph and plot temperature v. mass of the stars in the table.

Apply It

6. All of the stars on your graph are main-sequence stars. What is the relationship between the color and the temperature of a main-sequence star?

7. What is the relationship between the mass and the temperature of a main-sequence star? How are color and mass related?

8. 🗝 **Key Concept** Which star would be the most likely to eventually form a black hole? Why?

Star	Color	Temperature (K)	Mass in solar masses
Sun	Yellow	5,700	1
Alnilam	Blue-white	27,000	40
Altair	White	8,000	1.9
Alpha Centauri A	Yellow	6,000	1.08
Alpha Centauri B	Orange	4,370	0.68
Barnard's Star	Red	3,100	0.1
Epsilon Eridani	Orange	4,830	0.78
Hadar	Blue-white	25,500	10.5
Proxima Centauri	Red	3,000	0.12
Regulus	White	11,000	8
Sirius A	White	9,500	2.6
Spica	Blue-white	22,000	10.5
Vega	White	9,900	3

Lesson 3 EXTEND

Lesson 4

Reading Guide

Key Concepts
ESSENTIAL QUESTIONS

- What are the major types of galaxies?
- What is the Milky Way, and how is it related to the solar system?
- What is the Big Bang theory?

Vocabulary
galaxy p. 825
dark matter p. 825
Big Bang theory p. 830
Doppler shift p. 830

g Multilingual eGlossary

Galaxies and the Universe

Inquiry Disk in Space?

Yes, this is the disk of a galaxy—a huge collection of stars. You see this galaxy on its edge. If you were to look down on it from above, it would look like a two-armed spiral. Do you think all galaxies are shaped like spirals? What about the galaxy you live in?

Inquiry Launch Lab
20 minutes

Does the universe move?

Scientists think the universe is expanding. What does that mean? Are stars and galaxies moving away from each other? Is the universe moving?

1. Read and complete a lab safety form.
2. Copy the table at right into your Science Journal.
3. Use a **marker** to make three dots 5–7 cm apart on one side of a **large round balloon.** Label the dots A, B, and C. The dots represent galaxies.
4. Blow up the balloon to a diameter of about 8 cm. Hold the balloon closed as your partner uses a **measuring tape** to measure the distance between each galaxy on the balloon's surface. Record the distances on the table.
5. Repeat step 4 two more times, blowing up the balloon a little more each time.

Balloon size	A–B (cm)	B–C (cm)	A–C (cm)
Small			
Medium			
Large			

Think About This
1. What happened to the distances between galaxies as the balloon expanded?
2. If you were standing in one of the galaxies, what would you observe about the other galaxies?
3. **Key Concept** If the balloon were a model of the universe, what do you think might have caused galaxies to move in this way?

Galaxies

Most people live in towns or cities where houses are close together. Not many houses are found in the wilderness. Similarly, most stars exist in galaxies. **Galaxies** *are huge collections of stars.* The universe contains hundreds of billions of galaxies, and each galaxy can contain hundreds of billions of stars.

✓ **Reading Check** What are galaxies?

Dark Matter

Gravity holds stars together. Gravity also holds galaxies together. When astronomers examine how galaxies, such as those in **Figure 14,** rotate and gravitationally interact, they find that most of the matter in galaxies is invisible. *Matter that emits no light at any wavelength is* **dark matter.** Scientists hypothesize that more than 90 percent of the universe's mass is dark matter. Scientists do not fully understand dark matter. They do not know what composes it.

Figure 14 By examining interacting galaxies such as these, astronomers hypothesize that most mass in the universe is invisible dark matter.

Types of Galaxies

There are three major types of galaxies: spiral, elliptical, and irregular. Table 2 gives a brief description of each type.

 Key Concept Check What are the major types of galaxies?

Table 2 Types of Galaxies

Spiral Galaxies
The stars, gas, and dust in a spiral galaxy exist in spiral arms that begin at a central disk. Some spiral arms are long and symmetrical; others are short and stubby. Spiral galaxies are thicker near the center, a region called the central bulge. A spherical halo of globular clusters and older, redder stars surrounds the disk. NGC 5679, shown here, contains a pair of spiral galaxies.

Elliptical Galaxies
Unlike spiral galaxies, elliptical galaxies do not have internal structure. Some are spheres, like basketballs, while others resemble footballs. Elliptical galaxies have higher percentages of old, red stars than spiral galaxies do. They contain little or no gas and dust. Scientists suspect that many elliptical galaxies form by the gravitational merging of two or more spiral galaxies. The elliptical galaxy pictured here is NGC 5982, part of the Draco Group.

Irregular Galaxies
Irregular galaxies are oddly shaped. Many form from the gravitational pull of neighboring galaxies. Irregular galaxies contain many young stars and have areas of intense star formation. Shown here is the irregular galaxy NGC 1427A.

Inquiry MiniLab

20 minutes

Can you identify a galaxy?

The *Hubble Space Telescope,* shown below, is an orbiting telescope that gives astronomers clear pictures of the night sky. What kinds of galaxies can you see in pictures taken by the *Hubble Telescope*?

1. Study each image on the **Hubble Space Telescope image sheet.** For each image, identify at least two galaxies. Are they spiral, elliptical, or irregular? Write your observations in your Science Journal, labeled with the letter of the image.

Analyze and Conclude

1. **Draw Conclusions** Why are some galaxies easier to identify than others?
2. **Infer** What interactions do you see among some of the galaxies?
3. **Key Concept** Do you think the shapes of galaxies can change over time? Why or why not?

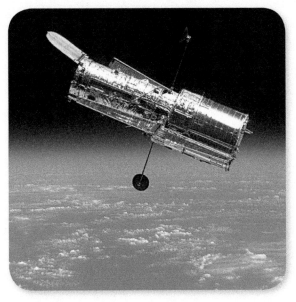

Groups of Galaxies

Galaxies are not distributed evenly in the universe. Gravity holds them together in groups called clusters. Some clusters of galaxies are enormous. The Virgo Cluster is 60 million light-years from Earth. It contains about 2,000 galaxies. Most clusters exist in even larger structures called superclusters. Between superclusters are voids, which are regions of nearly empty space. Scientists hypothesize that the large-scale structure of the universe resembles a sponge.

Reading Check What holds clusters of galaxies together?

The Milky Way

The solar system is in the Milky Way, a spiral galaxy that contains gas, dust, and almost 200 billion stars. The Milky Way is a member of the Local Group, a cluster of about 30 galaxies. Scientists expect the Milky Way will begin to merge with the Andromeda Galaxy, the largest galaxy in the Local Group, in about 3 billion years. Because stars are far apart in galaxies, it is not likely that many stars will actually collide during this event.

Where is Earth in the Milky Way? **Figure 15** on the next two pages shows an artist's drawing of the Milky Way and Earth's place in it.

FOLDABLES

Make a horizontal single-tab matchbook. Label it as shown. Use it to describe the contents of the Milky Way.

Milky Way
Galaxy

WORD ORIGIN

galaxy
from Greek *galactos*, means "milk"

The Milky Way

Figure 15 The Milky Way is shown here in two separate views, from the top (left page) and on edge (right page). Because Earth is located inside the disk of the Milky Way, people cannot see beyond the central bulge to the other side.

Key Concept Check Where is Earth in the Milky Way?

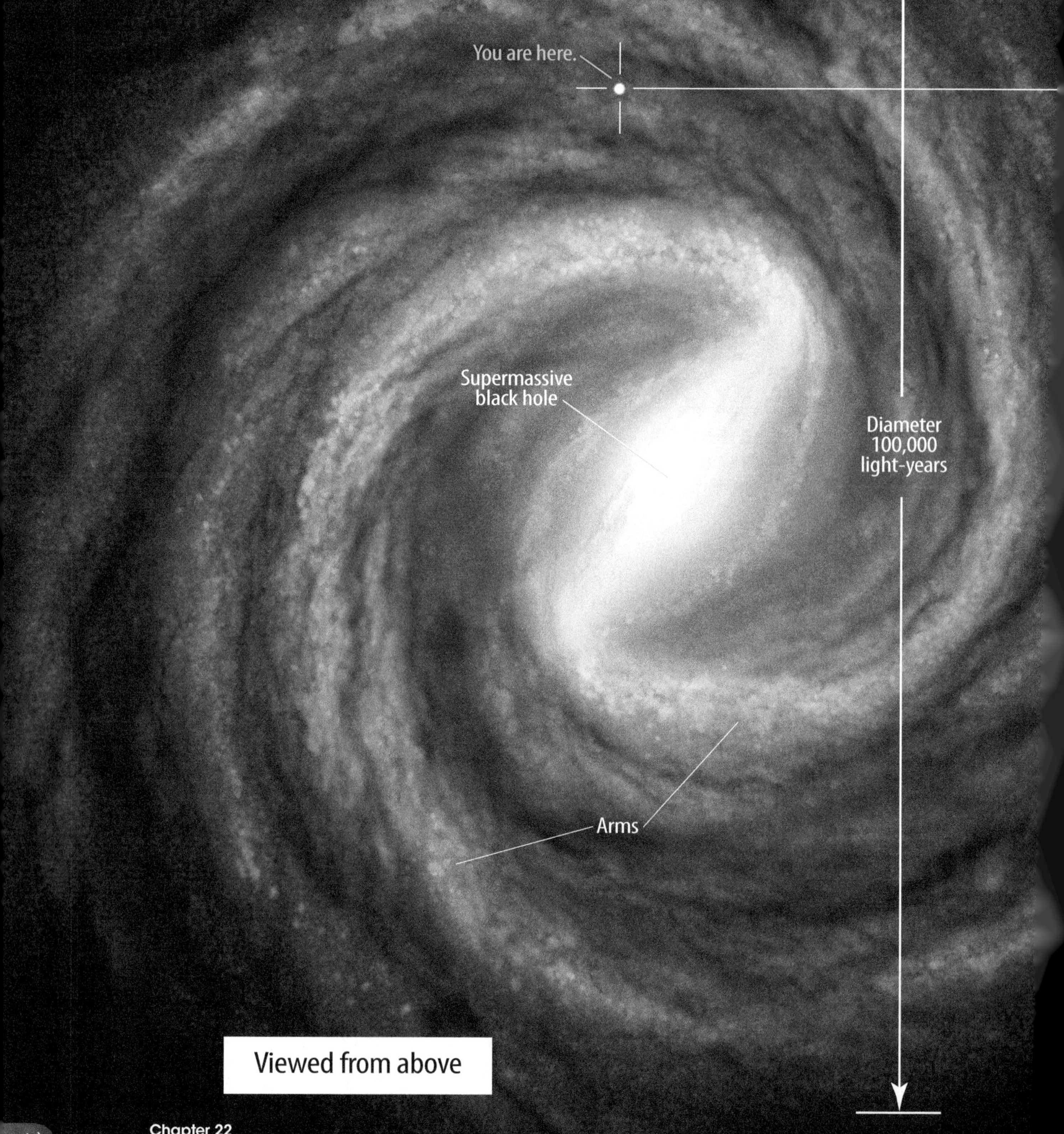

You are here.

Supermassive black hole

Diameter 100,000 light-years

Arms

Viewed from above

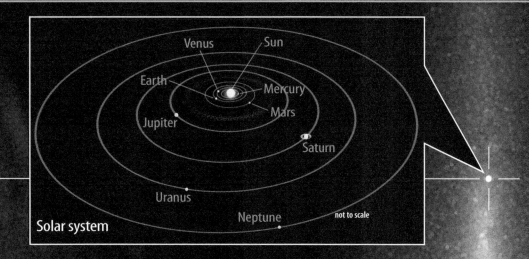

Solar system (not to scale)

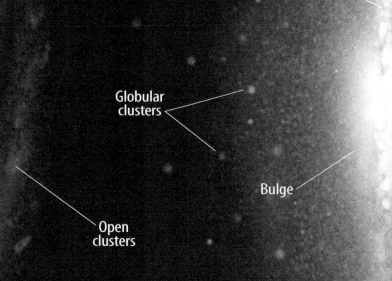

Viewed on edge

The Big Bang Theory

When astronomers look into space, they look back in time. Is there a beginning to time? According to the **Big Bang theory,** *the universe began from one point billions of years ago and has been expanding ever since.*

 Key Concept Check What is the Big Bang theory?

Origin and Expansion of the Universe

Most scientists agree that the universe is 13 – 14 billion years old. When the universe began, it was dense and hot—so hot that even atoms didn't exist. After a few hundred thousand years, the universe cooled enough for atoms to form. Eventually, stars formed, and gravity pulled them into galaxies.

As the universe expands, space stretches and galaxies move away from one another. The same thing happens in a loaf of unbaked raisin bread. As the dough rises, the raisins move apart. Scientists observe how space stretches by measuring the speed at which galaxies move away from Earth. As the galaxies move away, their wavelengths lengthen and stretch out. How does light stretch?

Doppler Shift

You have probably heard the siren of a speeding police car. As **Figure 16** illustrates, when the car moves toward you, the sound waves compress. As the car moves away, the sound waves spread out. Similarly, when visible light travels toward you, its wavelength compresses. When light travels away from you, its wavelength stretches out. It shifts to the red end of the visual light portion of the electromagnetic spectrum. *The shift to a different wavelength is called the* **Doppler shift.** Because the universe is expanding, light from galaxies is red-shifted. The more distant a galaxy is, the faster it moves away from Earth, and the more it is red-shifted.

Dark Energy

Will the universe expand forever? Or will gravity cause the universe to contract? Scientists have observed that galaxies are moving away from Earth faster over time. To explain this, they suggest a force called dark energy is pushing the galaxies apart.

Dark energy, like dark matter, is an active area of research. There is still much to learn about the universe and all it contains.

Doppler Shift

Figure 16 The sound waves from an approaching police car are compressed. As the car speeds away, the sound waves are stretched out. Similarly, when an object is moving away, its light is stretched out. The light's wavelength shifts toward a longer wavelength.

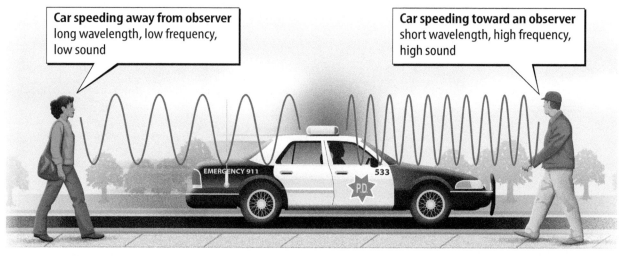

Lesson 4 Review

 Assessment   Online Quiz

Visual Summary

By studying interacting galaxies, scientists have determined that most mass in the universe is dark matter.

The Sun is one of billions of stars in the Milky Way.

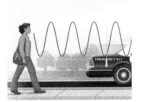
When an object moves away, its light stretches out, just as a siren's sound waves stretch out as the siren moves away.

FOLDABLES

Use your lesson Foldable to review the lesson. Save your Foldable for the project at the end of the chapter.

What do you think NOW?

You first read the statements below at the beginning of the chapter.

7. Most of the mass in the universe is in stars.

8. The Big Bang theory is an explanation of the beginning of the universe.

Did you change your mind about whether you agree or disagree with the statements? Rewrite any false statements to make them true.

Use Vocabulary

1 Stars exist in huge collections called _____.

2 **Use the term** *dark matter* in a sentence.

3 **Define** the *Big Bang theory*.

Understand Key Concepts

4 Which is NOT a major galaxy type?
   A. dark   C. irregular
   B. elliptical   D. spiral

5 **Identify** Sketch the Milky Way, and identify the location of the solar system.

6 **Explain** how scientists know the universe is expanding.

Interpret Graphics

7 **Identify** Sketch the Milky Way, shown below. Identify the bulge, the halo, and the disk.

8 **Organize Information** Copy and fill in the graphic organizer below. List the three major types of galaxies and some characteristics of each.

Galaxy Type	Characteristics

Critical Thinking

9 **Assess** the role of gravity in the structure of the universe.

10 **Predict** what the solar system and the universe might be like in 10 billion years.

Lesson 4
EVALUATE

Inquiry Lab

3 class periods

Materials

paper

colored pencils

astronomy magazines

string

glue

scissors

Safety

Describe a Trip Through Space

Imagine you could travel through space at speeds even faster than light. Based on what you have learned in this chapter, where would you choose to go? What would you like to see? What would it be like to move through the Milky Way and out into distant galaxies? Would you travel with anyone or meet any characters? Write a book describing your trip through space.

Question

Where will your trip take you, and how will you describe it? How can you write a fictional, but scientifically accurate, story about your trip? Will you make a picture book? If so, will you sketch your own pictures, use diagrams or photographs, or both? Will your book be mostly words, or will it be like a graphic novel? How can you draw your readers into the story?

Procedure

1. In your Science Journal, write ideas about where your trip will take you, how you will travel, what will happen along the way, and who or what you might meet.
2. Draw a graphic organizer, such as the one below, in your Science Journal. Use it to help you organize your ideas.
3. Write an outline of your story. Use it to guide you as you write the story.
4. List things you will need to research, pictures you will need to find or draw, and any other materials you will need. How will you bind your book? Will you make more than one copy?

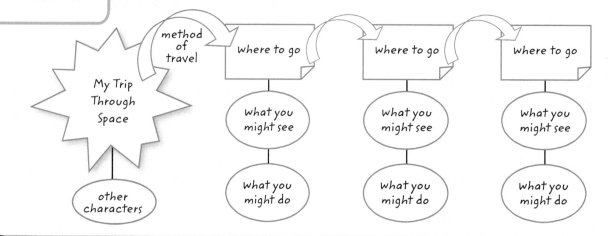

5. Write your book. Add pictures or illustrations. Bind the pages together into book form.
6. Have a friend read your book and tell you if you succeeded in telling your story in an engaging way. What suggestions does your friend have for improvement?
7. Revise and improve the book based on your friend's suggestions.

Analyze and Conclude

8. **Research Information** What new information did you learn as you did research for your book?
9. **Calculate** how many light-years you traveled from Earth.
10. **Draw Conclusions** How would your story be limited if you could only travel at the speed of light?
11. **The Big Idea** How does your story help people understand the size of the universe, what it contains, and how gravity affects it?

Communicate Your Results

You may wish to make a copy of your book and give it to the school library or add it to a library of books in your classroom.

Combine your book with books written by other students in your class to make an almanac of the universe. Add pages that give statistics and other interesting facts about the universe.

Lab Tips

☑ Think about your audience as you plan your book. Are you writing it for young children or for students your own age? What kinds of books do you and your friends enjoy reading?

☑ What metaphors or other kinds of figurative language can you add to your writing that will draw readers into your story?

Remember to use scientific methods.

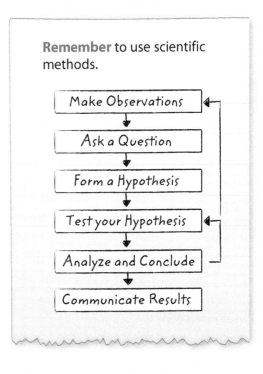

Chapter 22 Study Guide

THE BIG IDEA The universe is made up of stars, gas, and dust, as well as invisible dark matter. Material in the universe is not randomly arranged, but is pulled by gravity into galaxies.

Key Concepts Summary

Lesson 1: The View from Earth
- The sky is divided into 88 constellations.
- Astronomers learn about the energy, distance, temperature, and composition of stars by studying their light.
- Astronomers measure distances in space in **astronomical units** and in **light-years**. They measure star brightness as **apparent magnitude** and as **luminosity**.

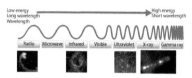

Lesson 2: The Sun and Other Stars
- **Stars** shine because of **nuclear fusion** in their cores.
- Stars have a layered structure—they conduct energy through their **radiative zones** and their **convection zones** and release the energy at their **photospheres.**
- Sunspots, prominences, flares, and coronal mass ejections are temporary phenomena on the Sun.
- Astronomers classify stars by their temperatures and luminosities.

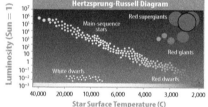

Lesson 3: Evolution of Stars
- Stars are born in clouds of gas and dust called **nebulae.**
- What happens to a star when it leaves the main sequence depends on its mass.
- Matter is recycled in the planetary nebulae of **white dwarfs** and the remnants of **supernovae.**

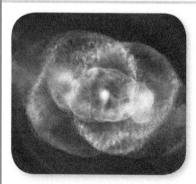

Lesson 4 Galaxies and the Universe
- The three major types of **galaxies** are spiral, elliptical, and irregular.
- The Milky Way is the spiral galaxy that contains the solar system.
- The **Big Bang theory** explains the origin of the universe.

Vocabulary

spectroscope p. 803
astronomical unit p. 804
light-year p. 804
apparent magnitude p. 805
luminosity p. 805

nuclear fusion p. 809
star p. 809
radiative zone p. 810
convection zone p. 810
photosphere p. 810
chromosphere p. 810
corona p. 810
Hertzsprung-Russell diagram p. 813

nebula p. 817
white dwarf p. 819
supernova p. 819
neutron star p. 820
black hole p. 820

galaxy p. 825
dark matter p. 825
Big Bang theory p. 830
Doppler shift p. 830

Study Guide

Review
- Personal Tutor
- Vocabulary eGames
- Vocabulary eFlashcards

FOLDABLES Chapter Project

Assemble your lesson Foldables as shown to make a Chapter Project. Use the project to review what you have learned in this chapter.

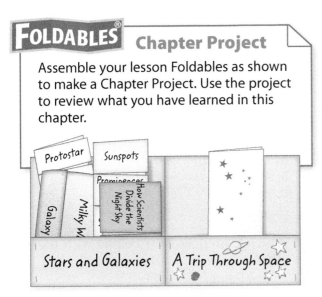

Use Vocabulary

1. **Explain** how nebulae are related to stars.
2. **Define** *Doppler shift*.
3. **Compare** neutron stars and black holes.
4. **Explain** the role of white dwarfs in recycling matter.
5. **Distinguish** between an astronomical unit and a light-year.
6. How does a convection zone transfer energy?
7. **Use the term** *dark matter* in a sentence.
8. On what diagram would you find a plot of stellar luminosity v. temperature?
9. **Compare** photosphere and corona.

 Interactive Concept Map

Link Vocabulary and Key Concepts

Copy this concept map, and then use vocabulary terms from the previous page to complete the concept map.

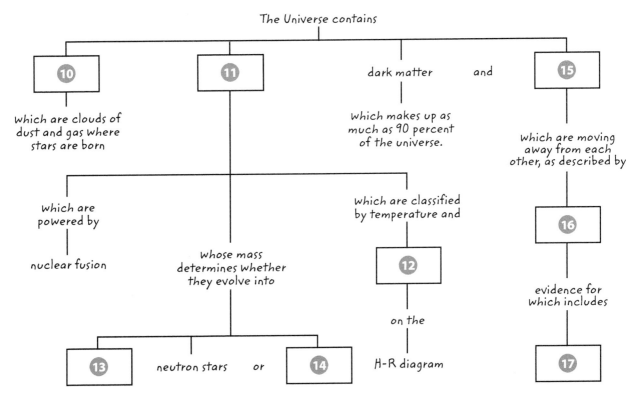

Chapter 22 Study Guide • **835**

Chapter 22 Review

Understand Key Concepts

1. Scientists divide the sky into
   A. astronomical units.
   B. clusters.
   C. constellations.
   D. light-years.

2. Which part of the Sun is marked with an X on the diagram below?

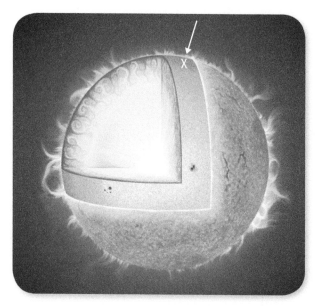

   A. convection zone
   B. corona
   C. photosphere
   D. radiative zone

3. Which might change, depending on the distance to a star?
   A. absolute magnitude
   B. apparent magnitude
   C. composition
   D. luminosity

4. Which is the average distance between Earth and the Sun?
   A. 1 AU
   B. 1 km
   C. 1 light-year
   D. 1 magnitude

5. Which is most important in determining the fate of a star?
   A. the star's color
   B. the star's distance
   C. the star's mass
   D. the star's temperature

6. What star along the main sequence will likely end in a supernova?
   A. blue-white
   B. orange
   C. red
   D. yellow

7. Which term does NOT belong with the others?
   A. black hole
   B. neutron star
   C. red dwarf
   D. supernova

8. What does the Big Bang theory state?
   A. The universe is ageless.
   B. The universe is collapsing.
   C. The universe is expanding.
   D. The universe is infinite.

9. Which type of galaxy is illustrated below?

   A. elliptical
   B. irregular
   C. peculiar
   D. spiral

836 • Chapter 22 Review

Chapter Review

Assessment — Online Test Practice

Critical Thinking

10 Explain how energy is released in a star.

11 Assess how the invention of the telescope changed people's views of the universe.

12 Imagine you are asked to classify 10,000 stars. Which properties would you measure?

13 Deduce why supernovae are needed for life on Earth.

14 Predict how the Sun would be different if it were twice as massive.

15 Imagine that you are writing to a friend who lives in the Virgo Cluster of galaxies. What would you write as your return address? Be specific.

16 Interpret The figure below shows part of the solar system. Explain what is happening.

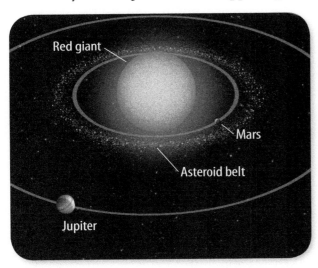

Writing in Science

17 Write You are a scientist being interviewed by a magazine on the topic of black holes. Write three questions an interviewer might ask, as well as your answers.

REVIEW THE BIG IDEA

18 What makes up the universe, and how does gravity affect the universe?

19 The photo below shows an image of the early universe obtained with the *Hubble Space Telescope*. Identify the objects you see. Make a list of other objects in the universe that you cannot see on this image.

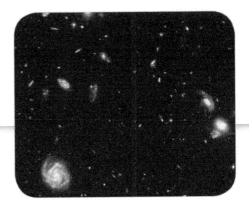

Math Skills

Review — Math Practice

Use Proportions

20 The Milky Way galaxy is about 100,000 light-years across. What is this distance in kilometers?

21 Astronomers sometimes use a distance unit called a parsec. One parsec is 3.3 light-years. What is the distance, in parsecs, of a nebula that is 82.5 light-years away?

22 The distance to the Orion nebula is about 390 parsecs. What is this distance in light-years?

Standardized Test Practice

Record your answers on the answer sheet provided by your teacher or on a sheet of paper.

Multiple Choice

1. Which characteristics can by studied by analyzing a star's spectrum?
   A absolute and apparent magnitudes
   B formation and evolution
   C movement and luminosity
   D temperature and composition

2. Which feature of the Sun appears in cycles of about 11 years?
   A coronal mass ejections
   B solar flares
   C solar wind
   D sunspots

Use the graph below to answer question 3.

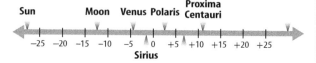

3. Which star on the graph has the greatest apparent magnitude?
   A Polaris
   B Proxima Centauri
   C Sirius
   D the Sun

4. Where in the Milky Way is the solar system located?
   A at the edge of the disk
   B inside a globular cluster
   C near the supermassive black hole
   D within the central bulge

Use the figure below to answer question 5.

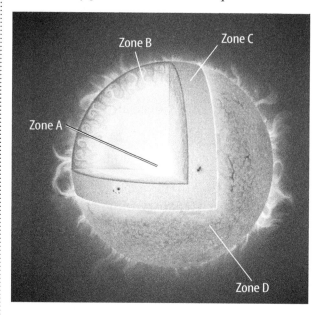

5. Which zone contains hot gas moving up toward the surface and cooler gas moving down toward the center of the Sun?
   A zone A
   B zone B
   C zone C
   D zone D

6. Which contains most of the mass of the universe?
   A black holes
   B dark matter
   C gas and dust
   D stars

7. Which stellar objects eventually form from the most massive stars?
   A black holes
   B diffuse nebulae
   C planetary nebulae
   D white dwarfs

Standardized Test Practice

Use the figure below to answer question 8.

8 Which is a characteristic for this type of galaxy?

A It contains no dust.

B It contains little gas.

C It contains many young stars.

D It contains mostly old stars.

9 Where do stars form?

A in black holes

B in constellations

C in nebulae

D in supernovae

10 What term describes the process that causes a star to shine?

A binary fission

B coronal mass ejection

C nuclear fusion

D stellar composition

11 What ancient star grouping do modern astronomers use to divide the sky into regions?

A astronomical unit

B constellation

C galaxy

D star cluster

Constructed Response

Use the diagram below to answer question 12.

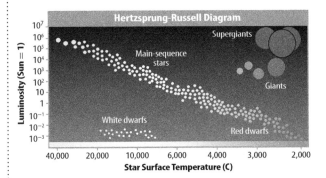

12 Use the information in the diagram above to describe red giants and white dwarfs based on their sizes, temperatures, and luminosities.

13 Describe the life cycle of a main-sequence star. What event causes the star to leave the main sequence?

14 How does the red shift of galaxies support the Big Bang theory?

15 Explain how planetary nebula recycle matter.

NEED EXTRA HELP?															
If You Missed Question...	1	2	3	4	5	6	7	8	9	10	11	12	13	14	15
Go to Lesson...	1	2	1	4	2	4	3	4	3	2	1	2	3	4	3

Student Resources

For Students and Parents/Guardians

These resources are designed to help you achieve success in science. You will find useful information on laboratory safety, math skills, and science skills. In addition, science reference materials are found in the Reference Handbook. You'll find the information you need to learn and sharpen your skills in these resources.

Table of Contents

Science Skill Handbook SR-2
Scientific Methods SR-2
- Identify a Question SR-2
- Gather and Organize Information SR-2
- Form a Hypothesis SR-5
- Test the Hypothesis SR-6
- Collect Data SR-6
- Analyze the Data SR-9
- Draw Conclustions SR-10
- Communicate SR-10

Safety Symbols SR-11
Safety in the Science Laboratory SR-12
- General Safety Rules SR-12
- Prevent Accidents SR-12
- Laboratory Work SR-13
- Emergencies SR-13

Math Skill Handbook SR-14
Math Review SR-14
- Use Fractions SR-14
- Use Ratios SR-17
- Use Decimals SR-17
- Use Proportions SR-18
- Use Percentages SR-19
- Solve One-Step Equations SR-19
- Use Statistics SR-20
- Use Geometry SR-21

Science Application SR-24
- Measure in SI SR-24
- Dimensional Analysis SR-24
- Precision and Significant Digits SR-26
- Scientific Notation SR-26
- Make and Use Graphs SR-27

Foldables Handbook SR-29

Reference Handbook SR-40
- Periodic Table of the Elements SR-40
- Topographic Map Symbols SR-42
- Rocks SR-43
- Minerals SR-44
- Weather Map Symbols SR-46

Glossary G-2

Index I-2

Credits C-2

Science Skill Handbook

Scientific Methods

Scientists use an orderly approach called the scientific method to solve problems. This includes organizing and recording data so others can understand them. Scientists use many variations in this method when they solve problems.

Identify a Question

The first step in a scientific investigation or experiment is to identify a question to be answered or a problem to be solved. For example, you might ask which gasoline is the most efficient.

Gather and Organize Information

After you have identified your question, begin gathering and organizing information. There are many ways to gather information, such as researching in a library, interviewing those knowledgeable about the subject, and testing and working in the laboratory and field. Fieldwork is investigations and observations done outside of a laboratory.

Researching Information Before moving in a new direction, it is important to gather the information that already is known about the subject. Start by asking yourself questions to determine exactly what you need to know. Then you will look for the information in various reference sources, like the student is doing in **Figure 1.** Some sources may include textbooks, encyclopedias, government documents, professional journals, science magazines, and the Internet. Always list the sources of your information.

Figure 1 The Internet can be a valuable research tool.

Evaluate Sources of Information Not all sources of information are reliable. You should evaluate all of your sources of information, and use only those you know to be dependable. For example, if you are researching ways to make homes more energy efficient, a site written by the U.S. Department of Energy would be more reliable than a site written by a company that is trying to sell a new type of weatherproofing material. Also, remember that research always is changing. Consult the most current resources available to you. For example, a 1985 resource about saving energy would not reflect the most recent findings.

Sometimes scientists use data that they did not collect themselves, or conclusions drawn by other researchers. This data must be evaluated carefully. Ask questions about how the data were obtained, if the investigation was carried out properly, and if it has been duplicated exactly with the same results. Would you reach the same conclusion from the data? Only when you have confidence in the data can you believe it is true and feel comfortable using it.

Interpret Scientific Illustrations As you research a topic in science, you will see drawings, diagrams, and photographs to help you understand what you read. Some illustrations are included to help you understand an idea that you can't see easily by yourself, like the tiny particles in an atom in **Figure 2.** A drawing helps many people to remember details more easily and provides examples that clarify difficult concepts or give additional information about the topic you are studying. Most illustrations have labels or a caption to identify or to provide more information.

Network Tree A type of concept map that not only shows a relationship, but how the concepts are related is a network tree, shown in **Figure 3.** In a network tree, the words are written in the ovals, while the description of the type of relationship is written across the connecting lines.

When constructing a network tree, write down the topic and all major topics on separate pieces of paper or notecards. Then arrange them in order from general to specific. Branch the related concepts from the major concept and describe the relationship on the connecting line. Continue to more specific concepts until finished.

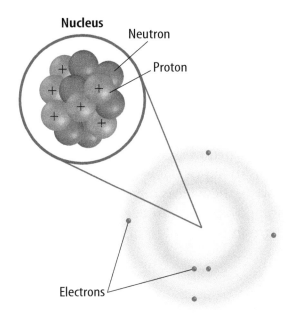

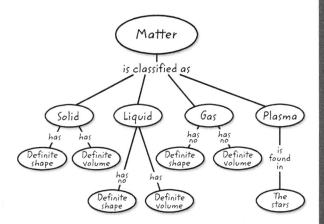

Figure 2 This drawing shows an atom of carbon with its six protons, six neutrons, and six electrons.

Figure 3 A network tree shows how concepts or objects are related.

Concept Maps One way to organize data is to draw a diagram that shows relationships among ideas (or concepts). A concept map can help make the meanings of ideas and terms more clear, and help you understand and remember what you are studying. Concept maps are useful for breaking large concepts down into smaller parts, making learning easier.

Events Chain Another type of concept map is an events chain. Sometimes called a flow chart, it models the order or sequence of items. An events chain can be used to describe a sequence of events, the steps in a procedure, or the stages of a process.

When making an events chain, first find the one event that starts the chain. This event is called the initiating event. Then, find the next event and continue until the outcome is reached, as shown in **Figure 4** on the next page.

Science Skill Handbook • **SR-3**

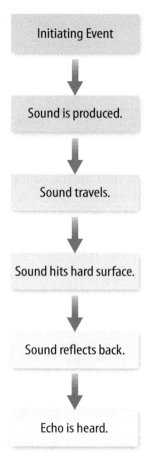

Figure 4 Events-chain concept maps show the order of steps in a process or event. This concept map shows how a sound makes an echo.

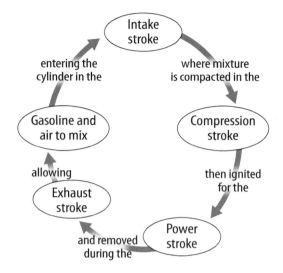

Figure 5 A cycle map shows events that occur in a cycle.

Cycle Map A specific type of events chain is a cycle map. It is used when the series of events do not produce a final outcome, but instead relate back to the beginning event, such as in **Figure 5**. Therefore, the cycle repeats itself.

To make a cycle map, first decide what event is the beginning event. This is also called the initiating event. Then list the next events in the order that they occur, with the last event relating back to the initiating event. Words can be written between the events that describe what happens from one event to the next. The number of events in a cycle map can vary, but usually contain three or more events.

Spider Map A type of concept map that you can use for brainstorming is the spider map. When you have a central idea, you might find that you have a jumble of ideas that relate to it but are not necessarily clearly related to each other. The spider map on sound in **Figure 6** shows that if you write these ideas outside the main concept, then you can begin to separate and group unrelated terms so they become more useful.

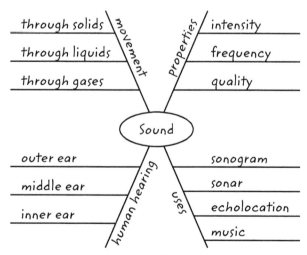

Figure 6 A spider map allows you to list ideas that relate to a central topic but not necessarily to one another.

SR-4 • Science Skill Handbook

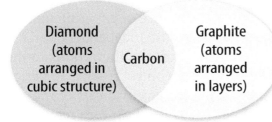

Figure 7 This Venn diagram compares and contrasts two substances made from carbon.

Venn Diagram To illustrate how two subjects compare and contrast you can use a Venn diagram. You can see the characteristics that the subjects have in common and those that they do not, shown in **Figure 7.**

To create a Venn diagram, draw two overlapping ovals that are big enough to write in. List the characteristics unique to one subject in one oval, and the characteristics of the other subject in the other oval. The characteristics in common are listed in the overlapping section.

Make and Use Tables One way to organize information so it is easier to understand is to use a table. Tables can contain numbers, words, or both.

To make a table, list the items to be compared in the first column and the characteristics to be compared in the first row. The title should clearly indicate the content of the table, and the column or row heads should be clear. Notice that in **Table 1** the units are included.

Table 1 Recyclables Collected During Week			
Day of Week	Paper (kg)	Aluminum (kg)	Glass (kg)
Monday	5.0	4.0	12.0
Wednesday	4.0	1.0	10.0
Friday	2.5	2.0	10.0

Make a Model One way to help you better understand the parts of a structure, the way a process works, or to show things too large or small for viewing is to make a model. For example, an atomic model made of a plastic-ball nucleus and chenille stem electron shells can help you visualize how the parts of an atom relate to each other. Other types of models can be devised on a computer or represented by equations.

Form a Hypothesis

A possible explanation based on previous knowledge and observations is called a hypothesis. After researching gasoline types and recalling previous experiences in your family's car you form a hypothesis—our car runs more efficiently because we use premium gasoline. To be valid, a hypothesis has to be something you can test by using an investigation.

Predict When you apply a hypothesis to a specific situation, you predict something about that situation. A prediction makes a statement in advance, based on prior observation, experience, or scientific reasoning. People use predictions to make everyday decisions. Scientists test predictions by performing investigations. Based on previous observations and experiences, you might form a prediction that cars are more efficient with premium gasoline. The prediction can be tested in an investigation.

Design an Experiment A scientist needs to make many decisions before beginning an investigation. Some of these include: how to carry out the investigation, what steps to follow, how to record the data, and how the investigation will answer the question. It also is important to address any safety concerns.

Science Skill Handbook • **SR-5**

Test the Hypothesis

Now that you have formed your hypothesis, you need to test it. Using an investigation, you will make observations and collect data, or information. This data might either support or not support your hypothesis. Scientists collect and organize data as numbers and descriptions.

Follow a Procedure In order to know what materials to use, as well as how and in what order to use them, you must follow a procedure. **Figure 8** shows a procedure you might follow to test your hypothesis.

Procedure	
Step 1	Use regular gasoline for two weeks.
Step 2	Record the number of kilometers between fill-ups and the amount of gasoline used.
Step 3	Switch to premium gasoline for two weeks.
Step 4	Record the number of kilometers between fill-ups and the amount of gasoline used.

Figure 8 A procedure tells you what to do step-by-step.

Identify and Manipulate Variables and Controls In any experiment, it is important to keep everything the same except for the item you are testing. The one factor you change is called the independent variable. The change that results is the dependent variable. Make sure you have only one independent variable, to assure yourself of the cause of the changes you observe in the dependent variable. For example, in your gasoline experiment the type of fuel is the independent variable. The dependent variable is the efficiency.

Many experiments also have a control—an individual instance or experimental subject for which the independent variable is not changed. You can then compare the test results to the control results. To design a control you can have two cars of the same type. The control car uses regular gasoline for four weeks. After you are done with the test, you can compare the experimental results to the control results.

Collect Data

Whether you are carrying out an investigation or a short observational experiment, you will collect data, as shown in **Figure 9**. Scientists collect data as numbers and descriptions and organize them in specific ways.

Observe Scientists observe items and events, then record what they see. When they use only words to describe an observation, it is called qualitative data. Scientists' observations also can describe how much there is of something. These observations use numbers, as well as words, in the description and are called quantitative data. For example, if a sample of the element gold is described as being "shiny and very dense" the data are qualitative. Quantitative data on this sample of gold might include "a mass of 30 g and a density of 19.3 g/cm^3."

Figure 9 Collecting data is one way to gather information directly.

Figure 10 Record data neatly and clearly so it is easy to understand.

When you make observations you should examine the entire object or situation first, and then look carefully for details. It is important to record observations accurately and completely. Always record your notes immediately as you make them, so you do not miss details or make a mistake when recording results from memory. Never put unidentified observations on scraps of paper. Instead they should be recorded in a notebook, like the one in **Figure 10.** Write your data neatly so you can easily read it later. At each point in the experiment, record your observations and label them. That way, you will not have to determine what the figures mean when you look at your notes later. Set up any tables that you will need to use ahead of time, so you can record any observations right away. Remember to avoid bias when collecting data by not including personal thoughts when you record observations. Record only what you observe.

Estimate Scientific work also involves estimating. To estimate is to make a judgment about the size or the number of something without measuring or counting. This is important when the number or size of an object or population is too large or too difficult to accurately count or measure.

Sample Scientists may use a sample or a portion of the total number as a type of estimation. To sample is to take a small, representative portion of the objects or organisms of a population for research. By making careful observations or manipulating variables within that portion of the group, information is discovered and conclusions are drawn that might apply to the whole population. A poorly chosen sample can be unrepresentative of the whole. If you were trying to determine the rainfall in an area, it would not be best to take a rainfall sample from under a tree.

Measure You use measurements every day. Scientists also take measurements when collecting data. When taking measurements, it is important to know how to use measuring tools properly. Accuracy also is important.

Length To measure length, the distance between two points, scientists use meters. Smaller measurements might be measured in centimeters or millimeters.

Length is measured using a metric ruler or meterstick. When using a metric ruler, line up the 0-cm mark with the end of the object being measured and read the number of the unit where the object ends. Look at the metric ruler shown in **Figure 11.** The centimeter lines are the long, numbered lines, and the shorter lines are millimeter lines. In this instance, the length would be 4.50 cm.

Figure 11 This metric ruler has centimeter and millimeter divisions.

Science Skill Handbook • **SR-7**

Mass The SI unit for mass is the kilogram (kg). Scientists can measure mass using units formed by adding metric prefixes to the unit gram (g), such as milligram (mg). To measure mass, you might use a triple-beam balance similar to the one shown in **Figure 12.** The balance has a pan on one side and a set of beams on the other side. Each beam has a rider that slides on the beam.

When using a triple-beam balance, place an object on the pan. Slide the largest rider along its beam until the pointer drops below zero. Then move it back one notch. Repeat the process for each rider proceeding from the larger to smaller until the pointer swings an equal distance above and below the zero point. Sum the masses on each beam to find the mass of the object. Move all riders back to zero when finished.

Instead of putting materials directly on the balance, scientists often take a tare of a container. A tare is the mass of a container into which objects or substances are placed for measuring their masses. To find the mass of objects or substances, find the mass of a clean container. Remove the container from the pan, and place the object or substances in the container. Find the mass of the container with the materials in it. Subtract the mass of the empty container from the mass of the filled container to find the mass of the materials you are using.

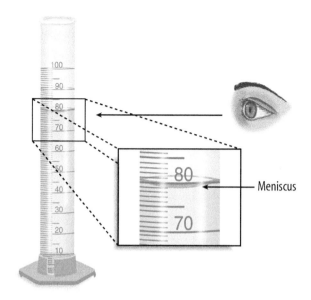

Figure 13 Graduated cylinders measure liquid volume.

Liquid Volume To measure liquids, the unit used is the liter. When a smaller unit is needed, scientists might use a milliliter. Because a milliliter takes up the volume of a cube measuring 1 cm on each side it also can be called a cubic centimeter ($cm^3 = cm \times cm \times cm$).

You can use beakers and graduated cylinders to measure liquid volume. A graduated cylinder, shown in **Figure 13,** is marked from bottom to top in milliliters. In lab, you might use a 10-mL graduated cylinder or a 100-mL graduated cylinder. When measuring liquids, notice that the liquid has a curved surface. Look at the surface at eye level, and measure the bottom of the curve. This is called the meniscus. The graduated cylinder in **Figure 13** contains 79.0 mL, or 79.0 cm^3, of a liquid.

Temperature Scientists often measure temperature using the Celsius scale. Pure water has a freezing point of 0°C and boiling point of 100°C. The unit of measurement is degrees Celsius. Two other scales often used are the Fahrenheit and Kelvin scales.

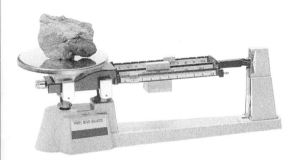

Figure 12 A triple-beam balance is used to determine the mass of an object.

SR-8 • Science Skill Handbook

Figure 14 A thermometer measures the temperature of an object.

Scientists use a thermometer to measure temperature. Most thermometers in a laboratory are glass tubes with a bulb at the bottom end containing a liquid such as colored alcohol. The liquid rises or falls with a change in temperature. To read a glass thermometer like the thermometer in **Figure 14,** rotate it slowly until a red line appears. Read the temperature where the red line ends.

Form Operational Definitions An operational definition defines an object by how it functions, works, or behaves. For example, when you are playing hide and seek and a tree is home base, you have created an operational definition for a tree.

Objects can have more than one operational definition. For example, a ruler can be defined as a tool that measures the length of an object (how it is used). It can also be a tool with a series of marks used as a standard when measuring (how it works).

Analyze the Data

To determine the meaning of your observations and investigation results, you will need to look for patterns in the data. Then you must think critically to determine what the data mean. Scientists use several approaches when they analyze the data they have collected and recorded. Each approach is useful for identifying specific patterns.

Interpret Data The word *interpret* means "to explain the meaning of something." When analyzing data from an experiment, try to find out what the data show. Identify the control group and the test group to see whether changes in the independent variable have had an effect. Look for differences in the dependent variable between the control and test groups.

Classify Sorting objects or events into groups based on common features is called classifying. When classifying, first observe the objects or events to be classified. Then select one feature that is shared by some members in the group, but not by all. Place those members that share that feature in a subgroup. You can classify members into smaller and smaller subgroups based on characteristics. Remember that when you classify, you are grouping objects or events for a purpose. Keep your purpose in mind as you select the features to form groups and subgroups.

Compare and Contrast Observations can be analyzed by noting the similarities and differences between two or more objects or events that you observe. When you look at objects or events to see how they are similar, you are comparing them. Contrasting is looking for differences in objects or events.

Science Skill Handbook • **SR-9**

Recognize Cause and Effect A cause is a reason for an action or condition. The effect is that action or condition. When two events happen together, it is not necessarily true that one event caused the other. Scientists must design a controlled investigation to recognize the exact cause and effect.

Draw Conclusions

When scientists have analyzed the data they collected, they proceed to draw conclusions about the data. These conclusions are sometimes stated in words similar to the hypothesis that you formed earlier. They may confirm a hypothesis, or lead you to a new hypothesis.

Infer Scientists often make inferences based on their observations. An inference is an attempt to explain observations or to indicate a cause. An inference is not a fact, but a logical conclusion that needs further investigation. For example, you may infer that a fire has caused smoke. Until you investigate, however, you do not know for sure.

Apply When you draw a conclusion, you must apply those conclusions to determine whether the data supports the hypothesis. If your data do not support your hypothesis, it does not mean that the hypothesis is wrong. It means only that the result of the investigation did not support the hypothesis. Maybe the experiment needs to be redesigned, or some of the initial observations on which the hypothesis was based were incomplete or biased. Perhaps more observation or research is needed to refine your hypothesis. A successful investigation does not always come out the way you originally predicted.

Avoid Bias Sometimes a scientific investigation involves making judgments. When you make a judgment, you form an opinion. It is important to be honest and not to allow any expectations of results to bias your judgments. This is important throughout the entire investigation, from researching to collecting data to drawing conclusions.

Communicate

The communication of ideas is an important part of the work of scientists. A discovery that is not reported will not advance the scientific community's understanding or knowledge. Communication among scientists also is important as a way of improving their investigations.

Scientists communicate in many ways, from writing articles in journals and magazines that explain their investigations and experiments, to announcing important discoveries on television and radio. Scientists also share ideas with colleagues on the Internet or present them as lectures, like the student is doing in **Figure 15**.

Figure 15 A student communicates to his peers about his investigation.

These safety symbols are used in laboratory and field investigations in this book to indicate possible hazards. Learn the meaning of each symbol and refer to this page often. *Remember to wash your hands thoroughly after completing lab procedures.*

PROTECTIVE EQUIPMENT Do not begin any lab without the proper protection equipment.

GOGGLES Proper eye protection must be worn when performing or observing science activities that involve items or conditions as listed below.	**APRON** Wear an approved apron when using substances that could stain, wet, or destroy cloth.	**SOAP** Wash hands with soap and water before removing goggles and after all lab activities.	**GLOVES** Wear gloves when working with biological materials, chemicals, animals, or materials that can stain or irritate hands.

LABORATORY HAZARDS

Symbols	Potential Hazards	Precaution	Response
DISPOSAL	contamination of classroom or environment due to improper disposal of materials such as chemicals and live specimens	• DO NOT dispose of hazardous materials in the sink or trash can. • Dispose of wastes as directed by your teacher.	• If hazardous materials are disposed of improperly, notify your teacher immediately.
EXTREME TEMPERATURE	skin burns due to extremely hot or cold materials such as hot glass, liquids, or metals; liquid nitrogen; dry ice	• Use proper protective equipment, such as hot mitts and/or tongs, when handling objects with extreme temperatures.	• If injury occurs, notify your teacher immediately.
SHARP OBJECTS	punctures or cuts from sharp objects such as razor blades, pins, scalpels, and broken glass	• Handle glassware carefully to avoid breakage. • Walk with sharp objects pointed downward, away from you and others.	• If broken glass or injury occurs, notify your teacher immediately.
ELECTRICAL	electric shock or skin burn due to improper grounding, short circuits, liquid spills, or exposed wires	• Check condition of wires and apparatus for fraying or uninsulated wires, and broken or cracked equipment. • Use only GFCI-protected outlets	• DO NOT attempt to fix electrical problems. Notify your teacher immediately.
CHEMICAL	skin irritation or burns, breathing difficulty, and/or poisoning due to touching, swallowing, or inhalation of chemicals such as acids, bases, bleach, metal compounds, iodine, poinsettias, pollen, ammonia, acetone, nail polish remover, heated chemicals, mothballs, and any other chemicals labeled or known to be dangerous	• Wear proper protective equipment such as goggles, apron, and gloves when using chemicals. • Ensure proper room ventilation or use a fume hood when using materials that produce fumes. • NEVER smell fumes directly. • NEVER taste or eat any material in the laboratory.	• If contact occurs, immediately flush affected area with water and notify your teacher. • If a spill occurs, leave the area immediately and notify your teacher.
FLAMMABLE	unexpected fire due to liquids or gases that ignite easily such as rubbing alcohol	• Avoid open flames, sparks, or heat when flammable liquids are present.	• If a fire occurs, leave the area immediately and notify your teacher.
OPEN FLAME	burns or fire due to open flame from matches, Bunsen burners, or burning materials	• Tie back loose hair and clothing. • Keep flame away from all materials. • Follow teacher instructions when lighting and extinguishing flames. • Use proper protection, such as hot mitts or tongs, when handling hot objects.	• If a fire occurs, leave the area immediately and notify your teacher.
ANIMAL SAFETY	injury to or from laboratory animals	• Wear proper protective equipment such as gloves, apron, and goggles when working with animals. • Wash hands after handling animals.	• If injury occurs, notify your teacher immediately.
BIOLOGICAL	infection or adverse reaction due to contact with organisms such as bacteria, fungi, and biological materials such as blood, animal or plant materials	• Wear proper protective equipment such as gloves, goggles, and apron when working with biological materials. • Avoid skin contact with an organism or any part of the organism. • Wash hands after handling organisms.	• If contact occurs, wash the affected area and notify your teacher immediately.
FUME	breathing difficulties from inhalation of fumes from substances such as ammonia, acetone, nail polish remover, heated chemicals, and mothballs	• Wear goggles, apron, and gloves. • Ensure proper room ventilation or use a fume hood when using substances that produce fumes. • NEVER smell fumes directly.	• If a spill occurs, leave area and notify your teacher immediately.
IRRITANT	irritation of skin, mucous membranes, or respiratory tract due to materials such as acids, bases, bleach, pollen, mothballs, steel wool, and potassium permanganate	• Wear goggles, apron, and gloves. • Wear a dust mask to protect against fine particles.	• If skin contact occurs, immediately flush the affected area with water and notify your teacher.
RADIOACTIVE	excessive exposure from alpha, beta, and gamma particles	• Remove gloves and wash hands with soap and water before removing remainder of protective equipment.	• If cracks or holes are found in the container, notify your teacher immediately.

Science Skill Handbook • **SR-11**

Safety in the Science Laboratory

Introduction to Science Safety

The science laboratory is a safe place to work if you follow standard safety procedures. Being responsible for your own safety helps to make the entire laboratory a safer place for everyone. When performing any lab, read and apply the caution statements and safety symbol listed at the beginning of the lab.

General Safety Rules

1. Complete the *Lab Safety Form* or other safety contract BEFORE starting any science lab.
2. Study the procedure. Ask your teacher any questions. Be sure you understand safety symbols shown on the page.
3. Notify your teacher about allergies or other health conditions that can affect your participation in a lab.
4. Learn and follow use and safety procedures for your equipment. If unsure, ask your teacher.

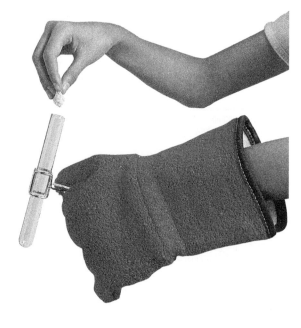

5. Never eat, drink, chew gum, apply cosmetics, or do any personal grooming in the lab. Never use lab glassware as food or drink containers. Keep your hands away from your face and mouth.
6. Know the location and proper use of the safety shower, eye wash, fire blanket, and fire alarm.

Prevent Accidents

1. Use the safety equipment provided to you. Goggles and a safety apron should be worn during investigations.
2. Do NOT use hair spray, mousse, or other flammable hair products. Tie back long hair and tie down loose clothing.
3. Do NOT wear sandals or other open-toed shoes in the lab.
4. Remove jewelry on hands and wrists. Loose jewelry, such as chains and long necklaces, should be removed to prevent them from getting caught in equipment.
5. Do not taste any substances or draw any material into a tube with your mouth.
6. Proper behavior is expected in the lab. Practical jokes and fooling around can lead to accidents and injury.
7. Keep your work area uncluttered.

Laboratory Work

1. Collect and carry all equipment and materials to your work area before beginning a lab.
2. Remain in your own work area unless given permission by your teacher to leave it.

3. Always slant test tubes away from yourself and others when heating them, adding substances to them, or rinsing them.
4. If instructed to smell a substance in a container, hold the container a short distance away and fan vapors toward your nose.
5. Do NOT substitute other chemicals/substances for those in the materials list unless instructed to do so by your teacher.
6. Do NOT take any materials or chemicals outside of the laboratory.
7. Stay out of storage areas unless instructed to be there and supervised by your teacher.

Laboratory Cleanup

1. Turn off all burners, water, and gas, and disconnect all electrical devices.
2. Clean all pieces of equipment and return all materials to their proper places.
3. Dispose of chemicals and other materials as directed by your teacher. Place broken glass and solid substances in the proper containers. Never discard materials in the sink.
4. Clean your work area.
5. Wash your hands with soap and water thoroughly BEFORE removing your goggles.

Emergencies

1. Report any fire, electrical shock, glassware breakage, spill, or injury, no matter how small, to your teacher immediately. Follow his or her instructions.
2. If your clothing should catch fire, STOP, DROP, and ROLL. If possible, smother it with the fire blanket or get under a safety shower. NEVER RUN.
3. If a fire should occur, turn off all gas and leave the room according to established procedures.
4. In most instances, your teacher will clean up spills. Do NOT attempt to clean up spills unless you are given permission and instructions to do so.
5. If chemicals come into contact with your eyes or skin, notify your teacher immediately. Use the eyewash, or flush your skin or eyes with large quantities of water.
6. The fire extinguisher and first-aid kit should only be used by your teacher unless it is an extreme emergency and you have been given permission.
7. If someone is injured or becomes ill, only a professional medical provider or someone certified in first aid should perform first-aid procedures.

Science Skill Handbook • **SR-13**

Math Skill Handbook

Math Review

Use Fractions

A fraction compares a part to a whole. In the fraction $\frac{2}{3}$, the 2 represents the part and is the numerator. The 3 represents the whole and is the denominator.

Reduce Fractions To reduce a fraction, you must find the largest factor that is common to both the numerator and the denominator, the greatest common factor (GCF). Divide both numbers by the GCF. The fraction has then been reduced, or it is in its simplest form.

Example

Twelve of the 20 chemicals in the science lab are in powder form. What fraction of the chemicals used in the lab are in powder form?

Step 1 Write the fraction.

$$\frac{\text{part}}{\text{whole}} = \frac{12}{20}$$

Step 2 To find the GCF of the numerator and denominator, list all of the factors of each number.

Factors of 12: 1, 2, 3, 4, 6, 12 (the numbers that divide evenly into 12)

Factors of 20: 1, 2, 4, 5, 10, 20 (the numbers that divide evenly into 20)

Step 3 List the common factors.

1, 2, 4

Step 4 Choose the greatest factor in the list. The GCF of 12 and 20 is 4.

Step 5 Divide the numerator and denominator by the GCF.

$$\frac{12 \div 4}{20 \div 4} = \frac{3}{5}$$

In the lab, $\frac{3}{5}$ of the chemicals are in powder form.

Practice Problem At an amusement park, 66 of 90 rides have a height restriction. What fraction of the rides, in its simplest form, has a height restriction?

Add and Subtract Fractions with Like Denominators To add or subtract fractions with the same denominator, add or subtract the numerators and write the sum or difference over the denominator. After finding the sum or difference, find the simplest form for your fraction.

Example 1

In the forest outside your house, $\frac{1}{8}$ of the animals are rabbits, $\frac{3}{8}$ are squirrels, and the remainder are birds and insects. How many are mammals?

Step 1 Add the numerators.

$$\frac{1}{8} + \frac{3}{8} = \frac{(1 + 3)}{8} = \frac{4}{8}$$

Step 2 Find the GCF.

$\frac{4}{8}$ (GCF, 4)

Step 3 Divide the numerator and denominator by the GCF.

$$\frac{4 \div 4}{8 \div 4} = \frac{1}{2}$$

$\frac{1}{2}$ of the animals are mammals.

Example 2

If $\frac{7}{16}$ of the Earth is covered by freshwater, and $\frac{1}{16}$ of that is in glaciers, how much freshwater is not frozen?

Step 1 Subtract the numerators.

$$\frac{7}{16} - \frac{1}{16} = \frac{(7 - 1)}{16} = \frac{6}{16}$$

Step 2 Find the GCF.

$\frac{6}{16}$ (GCF, 2)

Step 3 Divide the numerator and denominator by the GCF.

$$\frac{6 \div 2}{16 \div 2} = \frac{3}{8}$$

$\frac{3}{8}$ of the freshwater is not frozen.

Practice Problem A bicycle rider is riding at a rate of 15 km/h for $\frac{4}{9}$ of his ride, 10 km/h for $\frac{2}{9}$ of his ride, and 8 km/h for the remainder of the ride. How much of his ride is he riding at a rate greater than 8 km/h?

Add and Subtract Fractions with Unlike Denominators To add or subtract fractions with unlike denominators, first find the least common denominator (LCD). This is the smallest number that is a common multiple of both denominators. Rename each fraction with the LCD, and then add or subtract. Find the simplest form if necessary.

Example 1

A chemist makes a paste that is $\frac{1}{2}$ table salt (NaCl), $\frac{1}{3}$ sugar ($C_6H_{12}O_6$), and the remainder is water (H_2O). How much of the paste is a solid?

Step 1 Find the LCD of the fractions.

$\frac{1}{2} + \frac{1}{3}$ (LCD, 6)

Step 2 Rename each numerator and each denominator with the LCD.

Step 3 Add the numerators.

$\frac{3}{6} + \frac{2}{6} = \frac{(3+2)}{6} = \frac{5}{6}$

$\frac{5}{6}$ of the paste is a solid.

Example 2

The average precipitation in Grand Junction, CO, is $\frac{7}{10}$ inch in November, and $\frac{3}{5}$ inch in December. What is the total average precipitation?

Step 1 Find the LCD of the fractions.

$\frac{7}{10} + \frac{3}{5}$ (LCD, 10)

Step 2 Rename each numerator and each denominator with the LCD.

Step 3 Add the numerators.

$\frac{7}{10} + \frac{6}{10} = \frac{(7+6)}{10} = \frac{13}{10}$

$\frac{13}{10}$ inches total precipitation, or $1\frac{3}{10}$ inches.

Practice Problem On an electric bill, about $\frac{1}{8}$ of the energy is from solar energy and about $\frac{1}{10}$ is from wind power. How much of the total bill is from solar energy and wind power combined?

Example 3

In your body, $\frac{7}{10}$ of your muscle contractions are involuntary (cardiac and smooth muscle tissue). Smooth muscle makes $\frac{3}{15}$ of your muscle contractions. How many of your muscle contractions are made by cardiac muscle?

Step 1 Find the LCD of the fractions.

$\frac{7}{10} - \frac{3}{15}$ (LCD, 30)

Step 2 Rename each numerator and each denominator with the LCD.

$\frac{7 \times 3}{10 \times 3} = \frac{21}{30}$

$\frac{3 \times 2}{15 \times 2} = \frac{6}{30}$

Step 3 Subtract the numerators.

$\frac{21}{30} - \frac{6}{30} = \frac{(21-6)}{30} = \frac{15}{30}$

Step 4 Find the GCF.

$\frac{15}{30}$ (GCF, 15)

$\frac{1}{2}$

$\frac{1}{2}$ of all muscle contractions are cardiac muscle.

Example 4

Tony wants to make cookies that call for $\frac{3}{4}$ of a cup of flour, but he only has $\frac{1}{3}$ of a cup. How much more flour does he need?

Step 1 Find the LCD of the fractions.

$\frac{3}{4} - \frac{1}{3}$ (LCD, 12)

Step 2 Rename each numerator and each denominator with the LCD.

$\frac{3 \times 3}{4 \times 3} = \frac{9}{12}$

$\frac{1 \times 4}{3 \times 4} = \frac{4}{12}$

Step 3 Subtract the numerators.

$\frac{9}{12} - \frac{4}{12} = \frac{(9-4)}{12} = \frac{5}{12}$

$\frac{5}{12}$ of a cup of flour

Practice Problem Using the information provided to you in Example 3 above, determine how many muscle contractions are voluntary (skeletal muscle).

Multiply Fractions To multiply with fractions, multiply the numerators and multiply the denominators. Find the simplest form if necessary.

> **Example**
>
> Multiply $\frac{3}{5}$ by $\frac{1}{3}$.
>
> **Step 1** Multiply the numerators and denominators.
> $$\frac{3}{5} \times \frac{1}{3} = \frac{(3 \times 1)}{(5 \times 3)} = \frac{3}{15}$$
>
> **Step 2** Find the GCF.
> $$\frac{3}{15} \text{ (GCF, 3)}$$
>
> **Step 3** Divide the numerator and denominator by the GCF.
> $$\frac{3 \div 3}{15 \div 3} = \frac{1}{5}$$
>
> $\frac{3}{5}$ multiplied by $\frac{1}{3}$ is $\frac{1}{5}$.

Practice Problem Multiply $\frac{3}{14}$ by $\frac{5}{16}$.

Find a Reciprocal Two numbers whose product is 1 are called multiplicative inverses, or reciprocals.

> **Example**
>
> Find the reciprocal of $\frac{3}{8}$.
>
> **Step 1** Inverse the fraction by putting the denominator on top and the numerator on the bottom.
> $$\frac{8}{3}$$
>
> The reciprocal of $\frac{3}{8}$ is $\frac{8}{3}$.

Practice Problem Find the reciprocal of $\frac{4}{9}$.

Divide Fractions To divide one fraction by another fraction, multiply the dividend by the reciprocal of the divisor. Find the simplest form if necessary.

> **Example 1**
>
> Divide $\frac{1}{9}$ by $\frac{1}{3}$.
>
> **Step 1** Find the reciprocal of the divisor. The reciprocal of $\frac{1}{3}$ is $\frac{3}{1}$.
>
> **Step 2** Multiply the dividend by the reciprocal of the divisor.
> $$\frac{\frac{1}{9}}{\frac{1}{3}} = \frac{1}{9} \times \frac{3}{1} = \frac{(1 \times 3)}{(9 \times 1)} = \frac{3}{9}$$
>
> **Step 3** Find the GCF.
> $$\frac{3}{9} \text{ (GCF, 3)}$$
>
> **Step 4** Divide the numerator and denominator by the GCF.
> $$\frac{3 \div 3}{9 \div 3} = \frac{1}{3}$$
>
> $\frac{1}{9}$ divided by $\frac{1}{3}$ is $\frac{1}{3}$.

> **Example 2**
>
> Divide $\frac{3}{5}$ by $\frac{1}{4}$.
>
> **Step 1** Find the reciprocal of the divisor. The reciprocal of $\frac{1}{4}$ is $\frac{4}{1}$.
>
> **Step 2** Multiply the dividend by the reciprocal of the divisor.
> $$\frac{\frac{3}{5}}{\frac{1}{4}} = \frac{3}{5} \times \frac{4}{1} = \frac{(3 \times 4)}{(5 \times 1)} = \frac{12}{5}$$
>
> $\frac{3}{5}$ divided by $\frac{1}{4}$ is $\frac{12}{5}$ or $2\frac{2}{5}$.

Practice Problem Divide $\frac{3}{11}$ by $\frac{7}{10}$.

Use Ratios

When you compare two numbers by division, you are using a ratio. Ratios can be written 3 to 5, 3:5, or $\frac{3}{5}$. Ratios, like fractions, also can be written in simplest form.

Ratios can represent one type of probability, called odds. This is a ratio that compares the number of ways a certain outcome occurs to the number of possible outcomes. For example, if you flip a coin 100 times, what are the odds that it will come up heads? There are two possible outcomes, heads or tails, so the odds of coming up heads are 50:100. Another way to say this is that 50 out of 100 times the coin will come up heads. In its simplest form, the ratio is 1:2.

Example 1

A chemical solution contains 40 g of salt and 64 g of baking soda. What is the ratio of salt to baking soda as a fraction in simplest form?

Step 1 Write the ratio as a fraction.

$$\frac{\text{salt}}{\text{baking soda}} = \frac{40}{64}$$

Step 2 Express the fraction in simplest form. The GCF of 40 and 64 is 8.

$$\frac{40}{64} = \frac{40 \div 8}{64 \div 8} = \frac{5}{8}$$

The ratio of salt to baking soda in the sample is 5:8.

Example 2

Sean rolls a 6-sided die 6 times. What are the odds that the side with a 3 will show?

Step 1 Write the ratio as a fraction.

$$\frac{\text{number of sides with a 3}}{\text{number of possible sides}} = \frac{1}{6}$$

Step 2 Multiply by the number of attempts.

$\frac{1}{6} \times 6$ attempts $= \frac{6}{6}$ attempts $= 1$ attempt

1 attempt out of 6 will show a 3.

Practice Problem Two metal rods measure 100 cm and 144 cm in length. What is the ratio of their lengths in simplest form?

Use Decimals

A fraction with a denominator that is a power of ten can be written as a decimal. For example, 0.27 means $\frac{27}{100}$. The decimal point separates the ones place from the tenths place.

Any fraction can be written as a decimal using division. For example, the fraction $\frac{5}{8}$ can be written as a decimal by dividing 5 by 8. Written as a decimal, it is 0.625.

Add or Subtract Decimals When adding and subtracting decimals, line up the decimal points before carrying out the operation.

Example 1

Find the sum of 47.68 and 7.80.

Step 1 Line up the decimal places when you write the numbers.

```
  47.68
+  7.80
```

Step 2 Add the decimals.

```
  47.68
+  7.80
  55.48
```

The sum of 47.68 and 7.80 is 55.48.

Example 2

Find the difference of 42.17 and 15.85.

Step 1 Line up the decimal places when you write the number.

```
  42.17
- 15.85
```

Step 2 Subtract the decimals.

```
  42.17
- 15.85
  26.32
```

The difference of 42.17 and 15.85 is 26.32.

Practice Problem Find the sum of 1.245 and 3.842.

Math Skill Handbook • **SR-17**

Multiply Decimals To multiply decimals, multiply the numbers like numbers without decimal points. Count the decimal places in each factor. The product will have the same number of decimal places as the sum of the decimal places in the factors.

> **Example**
>
> Multiply 2.4 by 5.9.
>
> **Step 1** Multiply the factors like two whole numbers.
>
> $24 \times 59 = 1416$
>
> **Step 2** Find the sum of the number of decimal places in the factors. Each factor has one decimal place, for a sum of two decimal places.
>
> **Step 3** The product will have two decimal places.
>
> 14.16
>
> The product of 2.4 and 5.9 is 14.16.

Practice Problem Multiply 4.6 by 2.2.

Divide Decimals When dividing decimals, change the divisor to a whole number. To do this, multiply both the divisor and the dividend by the same power of ten. Then place the decimal point in the quotient directly above the decimal point in the dividend. Then divide as you do with whole numbers.

> **Example**
>
> Divide 8.84 by 3.4.
>
> **Step 1** Multiply both factors by 10.
>
> $3.4 \times 10 = 34, 8.84 \times 10 = 88.4$
>
> **Step 2** Divide 88.4 by 34.
>
> $$\begin{array}{r} 2.6 \\ 34\overline{)88.4} \\ -68 \\ \hline 204 \\ -204 \\ \hline 0 \end{array}$$
>
> 8.84 divided by 3.4 is 2.6.

Practice Problem Divide 75.6 by 3.6.

Use Proportions

An equation that shows that two ratios are equivalent is a proportion. The ratios $\frac{2}{4}$ and $\frac{5}{10}$ are equivalent, so they can be written as $\frac{2}{4} = \frac{5}{10}$. This equation is a proportion.

When two ratios form a proportion, the cross products are equal. To find the cross products in the proportion $\frac{2}{4} = \frac{5}{10}$, multiply the 2 and the 10, and the 4 and the 5. Therefore $2 \times 10 = 4 \times 5$, or $20 = 20$.

Because you know that both ratios are equal, you can use cross products to find a missing term in a proportion. This is known as solving the proportion.

> **Example**
>
> The heights of a tree and a pole are proportional to the lengths of their shadows. The tree casts a shadow of 24 m when a 6-m pole casts a shadow of 4 m. What is the height of the tree?
>
> **Step 1** Write a proportion.
>
> $$\frac{\text{height of tree}}{\text{height of pole}} = \frac{\text{length of tree's shadow}}{\text{length of pole's shadow}}$$
>
> **Step 2** Substitute the known values into the proportion. Let h represent the unknown value, the height of the tree.
>
> $\frac{h}{6} \times \frac{24}{4}$
>
> **Step 3** Find the cross products.
>
> $h \times 4 = 6 \times 24$
>
> **Step 4** Simplify the equation.
>
> $4h \times 144$
>
> **Step 5** Divide each side by 4.
>
> $\frac{4h}{4} \times \frac{144}{4}$
>
> $h = 36$
>
> The height of the tree is 36 m.

Practice Problem The ratios of the weights of two objects on the Moon and on Earth are in proportion. A rock weighing 3 N on the Moon weighs 18 N on Earth. How much would a rock that weighs 5 N on the Moon weigh on Earth?

Use Percentages

The word *percent* means "out of one hundred." It is a ratio that compares a number to 100. Suppose you read that 77 percent of Earth's surface is covered by water. That is the same as reading that the fraction of Earth's surface covered by water is $\frac{77}{100}$. To express a fraction as a percent, first find the equivalent decimal for the fraction. Then, multiply the decimal by 100 and add the percent symbol.

Example 1

Express $\frac{13}{20}$ as a percent.

Step 1 Find the equivalent decimal for the fraction.

$$\begin{array}{r} 0.65 \\ 20\overline{)13.00} \\ \underline{12\,0} \\ 1\,00 \\ \underline{1\,00} \\ 0 \end{array}$$

Step 2 Rewrite the fraction $\frac{13}{20}$ as 0.65.

Step 3 Multiply 0.65 by 100 and add the % symbol.

$0.65 \times 100 = 65 = 65\%$

So, $\frac{13}{20} = 65\%$.

This also can be solved as a proportion.

Example 2

Express $\frac{13}{20}$ as a percent.

Step 1 Write a proportion.

$\frac{13}{20} = \frac{x}{100}$

Step 2 Find the cross products.

$1300 = 20x$

Step 3 Divide each side by 20.

$\frac{1300}{20} = \frac{20x}{20}$

$65\% = x$

Practice Problem In one year, 73 of 365 days were rainy in one city. What percent of the days in that city were rainy?

Solve One-Step Equations

A statement that two expressions are equal is an equation. For example, $A = B$ is an equation that states that A is equal to B.

An equation is solved when a variable is replaced with a value that makes both sides of the equation equal. To make both sides equal the inverse operation is used. Addition and subtraction are inverses, and multiplication and division are inverses.

Example 1

Solve the equation $x - 10 = 35$.

Step 1 Find the solution by adding 10 to each side of the equation.

$$x - 10 = 35$$
$$x - 10 + 10 = 35 - 10$$
$$x = 45$$

Step 2 Check the solution.

$$x - 10 = 35$$
$$45 - 10 = 35$$
$$35 = 35$$

Both sides of the equation are equal, so $x = 45$.

Example 2

In the formula $a = bc$, find the value of c if $a = 20$ and $b = 2$.

Step 1 Rearrange the formula so the unknown value is by itself on one side of the equation by dividing both sides by b.

$$a = bc$$
$$\frac{a}{b} = \frac{bc}{b}$$
$$\frac{a}{b} = c$$

Step 2 Replace the variables a and b with the values that are given.

$$\frac{a}{b} = c$$
$$\frac{20}{2} = c$$
$$10 = c$$

Step 3 Check the solution.

$$a = bc$$
$$20 = 2 \times 10$$
$$20 = 20$$

Both sides of the equation are equal, so $c = 10$ is the solution when $a = 20$ and $b = 2$.

Practice Problem In the formula $h = gd$, find the value of d if $g = 12.3$ and $h = 17.4$.

Use Statistics

The branch of mathematics that deals with collecting, analyzing, and presenting data is statistics. In statistics, there are three common ways to summarize data with a single number—the mean, the median, and the mode.

The **mean** of a set of data is the arithmetic average. It is found by adding the numbers in the data set and dividing by the number of items in the set.

The **median** is the middle number in a set of data when the data are arranged in numerical order. If there were an even number of data points, the median would be the mean of the two middle numbers.

The **mode** of a set of data is the number or item that appears most often.

Another number that often is used to describe a set of data is the range. The **range** is the difference between the largest number and the smallest number in a set of data.

Example

The speeds (in m/s) for a race car during five different time trials are 39, 37, 44, 36, and 44.

To find the mean:

Step 1 Find the sum of the numbers.

$39 + 37 + 44 + 36 + 44 = 200$

Step 2 Divide the sum by the number of items, which is 5.

$200 \div 5 = 40$

The mean is 40 m/s.

To find the median:

Step 1 Arrange the measures from least to greatest.

36, 37, 39, 44, 44

Step 2 Determine the middle measure.

36, 37, 39, 44, 44

The median is 39 m/s.

To find the mode:

Step 1 Group the numbers that are the same together.

44, 44, 36, 37, 39

Step 2 Determine the number that occurs most in the set.

44, 44, 36, 37, 39

The mode is 44 m/s.

To find the range:

Step 1 Arrange the measures from greatest to least.

44, 44, 39, 37, 36

Step 2 Determine the greatest and least measures in the set.

44, 44, 39, 37, 36

Step 3 Find the difference between the greatest and least measures.

$44 - 36 = 8$

The range is 8 m/s.

Practice Problem Find the mean, median, mode, and range for the data set 8, 4, 12, 8, 11, 14, 16.

A **frequency table** shows how many times each piece of data occurs, usually in a survey. **Table 1** below shows the results of a student survey on favorite color.

Table 1 Student Color Choice		
Color	Tally	Frequency
red	IIII	4
blue	ͰͰͰͰ	5
black	II	2
green	III	3
purple	ͰͰͰͰ II	7
yellow	ͰͰͰͰ I	6

Based on the frequency table data, which color is the favorite?

Use Geometry

The branch of mathematics that deals with the measurement, properties, and relationships of points, lines, angles, surfaces, and solids is called geometry.

Perimeter The **perimeter** (P) is the distance around a geometric figure. To find the perimeter of a rectangle, add the length and width and multiply that sum by two, or $2(l + w)$. To find perimeters of irregular figures, add the length of the sides.

Example 1

Find the perimeter of a rectangle that is 3 m long and 5 m wide.

Step 1 You know that the perimeter is 2 times the sum of the width and length.

$P = 2(3 \text{ m} + 5 \text{ m})$

Step 2 Find the sum of the width and length.

$P = 2(8 \text{ m})$

Step 3 Multiply by 2.

$P = 16 \text{ m}$

The perimeter is 16 m.

Example 2

Find the perimeter of a shape with sides measuring 2 cm, 5 cm, 6 cm, 3 cm.

Step 1 You know that the perimeter is the sum of all the sides.

$P = 2 + 5 + 6 + 3$

Step 2 Find the sum of the sides.

$P = 2 + 5 + 6 + 3$

$P = 16$

The perimeter is 16 cm.

Practice Problem Find the perimeter of a rectangle with a length of 18 m and a width of 7 m.

Practice Problem Find the perimeter of a triangle measuring 1.6 cm by 2.4 cm by 2.4 cm.

Area of a Rectangle The **area** (A) is the number of square units needed to cover a surface. To find the area of a rectangle, multiply the length times the width, or $l \times w$. When finding area, the units also are multiplied. Area is given in square units.

Example

Find the area of a rectangle with a length of 1 cm and a width of 10 cm.

Step 1 You know that the area is the length multiplied by the width.

$A = (1 \text{ cm} \times 10 \text{ cm})$

Step 2 Multiply the length by the width. Also multiply the units.

$A = 10 \text{ cm}^2$

The area is 10 cm².

Practice Problem Find the area of a square whose sides measure 4 m.

Area of a Triangle To find the area of a triangle, use the formula:

$A = \frac{1}{2}(\text{base} \times \text{height})$

The base of a triangle can be any of its sides. The height is the perpendicular distance from a base to the opposite endpoint, or vertex.

Example

Find the area of a triangle with a base of 18 m and a height of 7 m.

Step 1 You know that the area is $\frac{1}{2}$ the base times the height.

$A = \frac{1}{2}(18 \text{ m} \times 7 \text{ m})$

Step 2 Multiply $\frac{1}{2}$ by the product of 18×7. Multiply the units.

$A = \frac{1}{2}(126 \text{ m}^2)$

$A = 63 \text{ m}^2$

The area is 63 m².

Practice Problem Find the area of a triangle with a base of 27 cm and a height of 17 cm.

Circumference of a Circle The **diameter** (d) of a circle is the distance across the circle through its center, and the **radius** (r) is the distance from the center to any point on the circle. The radius is half of the diameter. The distance around the circle is called the **circumference** (C). The formula for finding the circumference is:

$C = 2\pi r$ or $C = \pi d$

The circumference divided by the diameter is always equal to 3.1415926… This nonterminating and nonrepeating number is represented by the Greek letter π (pi). An approximation often used for π is 3.14.

Example 1

Find the circumference of a circle with a radius of 3 m.

Step 1 You know the formula for the circumference is 2 times the radius times π.

$C = 2\pi(3)$

Step 2 Multiply 2 times the radius.

$C = 6\pi$

Step 3 Multiply by π.

$C \approx 19$ m

The circumference is about 19 m.

Example 2

Find the circumference of a circle with a diameter of 24.0 cm.

Step 1 You know the formula for the circumference is the diameter times π.

$C = \pi(24.0)$

Step 2 Multiply the diameter by π.

$C \approx 75.4$ cm

The circumference is about 75.4 cm.

Practice Problem Find the circumference of a circle with a radius of 19 cm.

Area of a Circle The formula for the area of a circle is: $A = \pi r^2$

Example 1

Find the area of a circle with a radius of 4.0 cm.

Step 1 $A = \pi(4.0)^2$

Step 2 Find the square of the radius.

$A = 16\pi$

Step 3 Multiply the square of the radius by π.

$A \approx 50$ cm^2

The area of the circle is about 50 cm^2.

Example 2

Find the area of a circle with a radius of 225 m.

Step 1 $A = \pi(225)^2$

Step 2 Find the square of the radius.

$A = 50625\pi$

Step 3 Multiply the square of the radius by π.

$A \approx 159043.1$

The area of the circle is about 159043.1 m^2.

Example 3

Find the area of a circle whose diameter is 20.0 mm.

Step 1 Remember that the radius is half of the diameter.

$A = \pi\left(\dfrac{20.0}{2}\right)^2$

Step 2 Find the radius.

$A = \pi(10.0)^2$

Step 3 Find the square of the radius.

$A = 100\pi$

Step 4 Multiply the square of the radius by π.

$A \approx 314$ mm^2

The area of the circle is about 314 mm^2.

Practice Problem Find the area of a circle with a radius of 16 m.

Volume The measure of space occupied by a solid is the **volume** (V). To find the volume of a rectangular solid multiply the length times width times height, or $V = l \times w \times h$. It is measured in cubic units, such as cubic centimeters (cm^3).

Example

Find the volume of a rectangular solid with a length of 2.0 m, a width of 4.0 m, and a height of 3.0 m.

Step 1 You know the formula for volume is the length times the width times the height.

$V = 2.0 \text{ m} \times 4.0 \text{ m} \times 3.0 \text{ m}$

Step 2 Multiply the length times the width times the height.

$V = 24 \text{ m}^3$

The volume is 24 m^3.

Practice Problem Find the volume of a rectangular solid that is 8 m long, 4 m wide, and 4 m high.

To find the volume of other solids, multiply the area of the base times the height.

Example 1

Find the volume of a solid that has a triangular base with a length of 8.0 m and a height of 7.0 m. The height of the entire solid is 15.0 m.

Step 1 You know that the base is a triangle, and the area of a triangle is $\frac{1}{2}$ the base times the height, and the volume is the area of the base times the height.

$V = \left[\frac{1}{2}(b \times h)\right] \times 15$

Step 2 Find the area of the base.

$V = \left[\frac{1}{2}(8 \times 7)\right] \times 15$

$V = \left(\frac{1}{2} \times 56\right) \times 15$

Step 3 Multiply the area of the base by the height of the solid.

$V = 28 \times 15$

$V = 420 \text{ m}^3$

The volume is 420 m^3.

Example 2

Find the volume of a cylinder that has a base with a radius of 12.0 cm, and a height of 21.0 cm.

Step 1 You know that the base is a circle, and the area of a circle is the square of the radius times π, and the volume is the area of the base times the height.

$V = (\pi r^2) \times 21$

$V = (\pi 12^2) \times 21$

Step 2 Find the area of the base.

$V = 144\pi \times 21$

$V = 452 \times 21$

Step 3 Multiply the area of the base by the height of the solid.

$V \approx 9{,}500 \text{ cm}^3$

The volume is about 9,500 cm^3.

Example 3

Find the volume of a cylinder that has a diameter of 15 mm and a height of 4.8 mm.

Step 1 You know that the base is a circle with an area equal to the square of the radius times π. The radius is one-half the diameter. The volume is the area of the base times the height.

$V = (\pi r^2) \times 4.8$

$V = \left[\pi\left(\frac{1}{2} \times 15\right)^2\right] \times 4.8$

$V = (\pi 7.5^2) \times 4.8$

Step 2 Find the area of the base.

$V = 56.25\pi \times 4.8$

$V \approx 176.71 \times 4.8$

Step 3 Multiply the area of the base by the height of the solid.

$V \approx 848.2$

The volume is about 848.2 mm^3.

Practice Problem Find the volume of a cylinder with a diameter of 7 cm in the base and a height of 16 cm.

Science Applications

Measure in SI

The metric system of measurement was developed in 1795. A modern form of the metric system, called the International System (SI), was adopted in 1960 and provides the standard measurements that all scientists around the world can understand.

The SI system is convenient because unit sizes vary by powers of 10. Prefixes are used to name units. Look at **Table 2** for some common SI prefixes and their meanings.

Table 2 Common SI Prefixes

Prefix	Symbol	Meaning	
kilo–	k	1,000	thousandth
hecto–	h	100	hundred
deka–	da	10	ten
deci–	d	0.1	tenth
centi–	c	0.01	hundreth
milli–	m	0.001	thousandth

Example

How many grams equal one kilogram?

Step 1 Find the prefix *kilo–* in **Table 2**.

Step 2 Using **Table 2**, determine the meaning of *kilo–*. According to the table, it means 1,000. When the prefix *kilo–* is added to a unit, it means that there are 1,000 of the units in a "kilounit."

Step 3 Apply the prefix to the units in the question. The units in the question are grams. There are 1,000 grams in a kilogram.

Practice Problem Is a milligram larger or smaller than a gram? How many of the smaller units equal one larger unit? What fraction of the larger unit does one smaller unit represent?

Dimensional Analysis

Convert SI Units In science, quantities such as length, mass, and time sometimes are measured using different units. A process called dimensional analysis can be used to change one unit of measure to another. This process involves multiplying your starting quantity and units by one or more conversion factors. A conversion factor is a ratio equal to one and can be made from any two equal quantities with different units. If 1,000 mL equal 1 L then two ratios can be made.

$$\frac{1,000 \text{ mL}}{1 \text{ L}} = \frac{1 \text{ L}}{1,000 \text{ mL}} = 1$$

One can convert between units in the SI system by using the equivalents in **Table 2** to make conversion factors.

Example

How many cm are in 4 m?

Step 1 Write conversion factors for the units given. From **Table 2**, you know that 100 cm = 1 m. The conversion factors are

$$\frac{100 \text{ cm}}{1 \text{ m}} \text{ and } \frac{1 \text{ m}}{100 \text{ cm}}$$

Step 2 Decide which conversion factor to use. Select the factor that has the units you are converting from (m) in the denominator and the units you are converting to (cm) in the numerator.

$$\frac{100 \text{ cm}}{1 \text{ m}}$$

Step 3 Multiply the starting quantity and units by the conversion factor. Cancel the starting units with the units in the denominator. There are 400 cm in 4 m.

$$4 \text{ m} = \frac{100 \text{ cm}}{1 \text{ m}} = 400 \text{ cm}$$

Practice Problem How many milligrams are in one kilogram? (Hint: You will need to use two conversion factors from **Table 2**.)

Table 3 Unit System Equivalents

Type of Measurement	Equivalent
Length	1 in = 2.54 cm 1 yd = 0.91 m 1 mi = 1.61 km
Mass and weight*	1 oz = 28.35 g 1 lb = 0.45 kg 1 ton (short) = 0.91 tonnes (metric tons) 1 lb = 4.45 N
Volume	1 in^3 = 16.39 cm^3 1 qt = 0.95 L 1 gal = 3.78 L
Area	1 in^2 = 6.45 cm^2 1 yd^2 = 0.83 m^2 1 mi^2 = 2.59 km^2 1 acre = 0.40 hectares
Temperature	°C = $\frac{(°F - 32)}{1.8}$ K = °C + 273

*Weight is measured in standard Earth gravity.

Convert Between Unit Systems Table 3 gives a list of equivalents that can be used to convert between English and SI units.

Example

If a meterstick has a length of 100 cm, how long is the meterstick in inches?

Step 1 Write the conversion factors for the units given. From **Table 3**, 1 in = 2.54 cm.

$$\frac{1 \text{ in}}{2.54 \text{ cm}} \text{ and } \frac{2.54 \text{ cm}}{1 \text{ in}}$$

Step 2 Determine which conversion factor to use. You are converting from cm to in. Use the conversion factor with cm on the bottom.

$$\frac{1 \text{ in}}{2.54 \text{ cm}}$$

Step 3 Multiply the starting quantity and units by the conversion factor. Cancel the starting units with the units in the denominator. Round your answer to the nearest tenth.

$$100 \text{ cm} \times \frac{1 \text{ in}}{2.54 \text{ cm}} = 39.37 \text{ in}$$

The meterstick is about 39.4 in long.

Practice Problem 1 A book has a mass of 5 lb. What is the mass of the book in kg?

Practice Problem 2 Use the equivalent for in and cm (1 in = 2.54 cm) to show how 1 in^3 ≈ 16.39 cm^3.

Math Skill Handbook • **SR-25**

Precision and Significant Digits

When you make a measurement, the value you record depends on the precision of the measuring instrument. This precision is represented by the number of significant digits recorded in the measurement. When counting the number of significant digits, all digits are counted except zeros at the end of a number with no decimal point such as 2,050, and zeros at the beginning of a decimal such as 0.03020. When adding or subtracting numbers with different precision, round the answer to the smallest number of decimal places of any number in the sum or difference. When multiplying or dividing, the answer is rounded to the smallest number of significant digits of any number being multiplied or divided.

Example

The lengths 5.28 and 5.2 are measured in meters. Find the sum of these lengths and record your answer using the correct number of significant digits.

Step 1 Find the sum.

```
   5.28 m    2 digits after the decimal
 + 5.2 m     1 digit after the decimal
  10.48 m
```

Step 2 Round to one digit after the decimal because the least number of digits after the decimal of the numbers being added is 1.

The sum is 10.5 m.

Practice Problem 1 How many significant digits are in the measurement 7,071,301 m? How many significant digits are in the measurement 0.003010 g?

Practice Problem 2 Multiply 5.28 and 5.2 using the rule for multiplying and dividing. Record the answer using the correct number of significant digits.

Scientific Notation

Many times numbers used in science are very small or very large. Because these numbers are difficult to work with scientists use scientific notation. To write numbers in scientific notation, move the decimal point until only one non-zero digit remains on the left. Then count the number of places you moved the decimal point and use that number as a power of ten. For example, the average distance from the Sun to Mars is 227,800,000,000 m. In scientific notation, this distance is 2.278×10^{11} m. Because you moved the decimal point to the left, the number is a positive power of ten.

The mass of an electron is about 0.000 000 000 000 000 000 000 000 000 911 kg. Expressed in scientific notation, this mass is 9.11×10^{-31} kg. Because the decimal point was moved to the right, the number is a negative power of ten.

Example

Earth is 149,600,000 km from the Sun. Express this in scientific notation.

Step 1 Move the decimal point until one non-zero digit remains on the left.

1.496 000 00

Step 2 Count the number of decimal places you have moved. In this case, eight.

Step 2 Show that number as a power of ten, 10^8.

Earth is 1.496×10^8 km from the Sun.

Practice Problem 1 How many significant digits are in 149,600,000 km? How many significant digits are in 1.496×10^8 km?

Practice Problem 2 Parts used in a high performance car must be measured to 7×10^{-6} m. Express this number as a decimal.

Practice Problem 3 A CD is spinning at 539 revolutions per minute. Express this number in scientific notation.

Make and Use Graphs

Data in tables can be displayed in a graph—a visual representation of data. Common graph types include line graphs, bar graphs, and circle graphs.

Line Graph A line graph shows a relationship between two variables that change continuously. The independent variable is changed and is plotted on the x-axis. The dependent variable is observed, and is plotted on the y-axis.

Example

Draw a line graph of the data below from a cyclist in a long-distance race.

Table 4 Bicycle Race Data

Time (h)	Distance (km)
0	0
1	8
2	16
3	24
4	32
5	40

Step 1 Determine the x-axis and y-axis variables. Time varies independently of distance and is plotted on the x-axis. Distance is dependent on time and is plotted on the y-axis.

Step 2 Determine the scale of each axis. The x-axis data ranges from 0 to 5. The y-axis data ranges from 0 to 50.

Step 3 Using graph paper, draw and label the axes. Include units in the labels.

Step 4 Draw a point at the intersection of the time value on the x-axis and corresponding distance value on the y-axis. Connect the points and label the graph with a title, as shown in **Figure 8**.

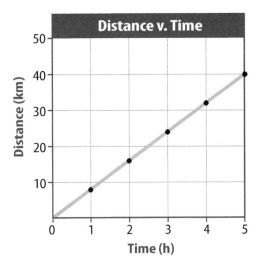

Figure 8 This line graph shows the relationship between distance and time during a bicycle ride.

Practice Problem A puppy's shoulder height is measured during the first year of her life. The following measurements were collected: (3 mo, 52 cm), (6 mo, 72 cm), (9 mo, 83 cm), (12 mo, 86 cm). Graph this data.

Find a Slope The slope of a straight line is the ratio of the vertical change, rise, to the horizontal change, run.

$$\text{Slope} = \frac{\text{vertical change (rise)}}{\text{horizontal change (run)}} = \frac{\text{change in } y}{\text{change in } x}$$

Example

Find the slope of the graph in **Figure 8**.

Step 1 You know that the slope is the change in y divided by the change in x.

$$\text{Slope} = \frac{\text{change in } y}{\text{change in } x}$$

Step 2 Determine the data points you will be using. For a straight line, choose the two sets of points that are the farthest apart.

$$\text{Slope} = \frac{(40 - 0) \text{ km}}{(5 - 0) \text{ h}}$$

Step 3 Find the change in y and x.

$$\text{Slope} = \frac{40 \text{ km}}{5 \text{ h}}$$

Step 4 Divide the change in y by the change in x.

$$\text{Slope} = \frac{8 \text{ km}}{\text{h}}$$

The slope of the graph is 8 km/h.

Bar Graph To compare data that does not change continuously you might choose a bar graph. A bar graph uses bars to show the relationships between variables. The *x*-axis variable is divided into parts. The parts can be numbers such as years, or a category such as a type of animal. The *y*-axis is a number and increases continuously along the axis.

Example

A recycling center collects 4.0 kg of aluminum on Monday, 1.0 kg on Wednesday, and 2.0 kg on Friday. Create a bar graph of this data.

Step 1 Select the *x*-axis and *y*-axis variables. The measured numbers (the masses of aluminum) should be placed on the *y*-axis. The variable divided into parts (collection days) is placed on the *x*-axis.

Step 2 Create a graph grid like you would for a line graph. Include labels and units.

Step 3 For each measured number, draw a vertical bar above the *x*-axis value up to the *y*-axis value. For the first data point, draw a vertical bar above Monday up to 4.0 kg.

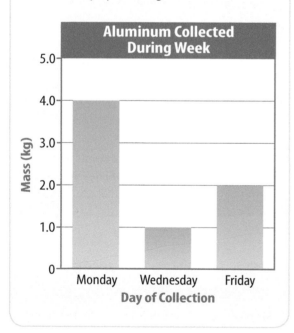

Practice Problem Draw a bar graph of the gases in air: 78% nitrogen, 21% oxygen, 1% other gases.

Circle Graph To display data as parts of a whole, you might use a circle graph. A circle graph is a circle divided into sections that represent the relative size of each piece of data. The entire circle represents 100%, half represents 50%, and so on.

Example

Air is made up of 78% nitrogen, 21% oxygen, and 1% other gases. Display the composition of air in a circle graph.

Step 1 Multiply each percent by 360° and divide by 100 to find the angle of each section in the circle.

$78\% \times \frac{360°}{100} = 280.8°$

$21\% \times \frac{360°}{100} = 75.6°$

$1\% \times \frac{360°}{100} = 3.6°$

Step 2 Use a compass to draw a circle and to mark the center of the circle. Draw a straight line from the center to the edge of the circle.

Step 3 Use a protractor and the angles you calculated to divide the circle into parts. Place the center of the protractor over the center of the circle and line the base of the protractor over the straight line.

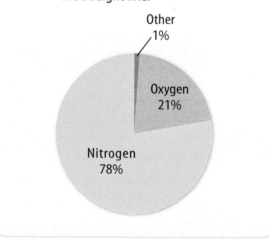

Practice Problem Draw a circle graph to represent the amount of aluminum collected during the week shown in the bar graph to the left.

Student Study Guides & Instructions
By Dinah Zike

1. You will find suggestions for Study Guides, also known as Foldables or books, in each chapter lesson and as a final project. Look at the end of the chapter to determine the project format and glue the Foldables in place as you progress through the chapter lessons.

2. Creating the Foldables or books is simple and easy to do by using copy paper, art paper, and internet printouts. Photocopies of maps, diagrams, or your own illustrations may also be used for some of the Foldables. Notebook paper is the most common source of material for study guides and 83% of all Foldables are created from it. When folded to make books, notebook paper Foldables easily fit into 11" × 17" or 12" × 18" chapter projects with space left over. Foldables made using photocopy paper are slightly larger and they fit into Projects, but snugly. Use the least amount of glue, tape, and staples needed to assemble the Foldables.

3. Seven of the Foldables can be made using either small or large paper. When 11" × 17" or 12" × 18" paper is used, these become projects for housing smaller Foldables. Project format boxes are located within the instructions to remind you of this option.

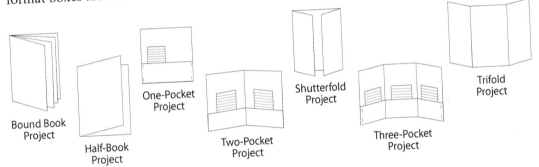

4. Use one-gallon self-locking plastic bags to store your projects. Place strips of two-inch clear tape along the left, long side of the bag and punch holes through the taped edge. Cut the bottom corners off the bag so it will not hold air. Store this Project Portfolio inside a three-hole binder. To store a large collection of project bags, use a giant laundry-soap box. Holes can be punched in some of the Foldable Projects so they can be stored in a three-hole binder without using a plastic bag. Punch holes in the pocket books before gluing or stapling the pocket.

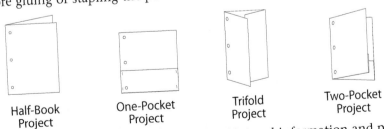

5. Maximize the use of the projects by collecting additional information and placing it on the back of the project and other unused spaces of the large Foldables.

Foldables Handbook • SR-29

Half-Book Foldable® By Dinah Zike

Step 1 Fold a sheet of notebook or copy paper in half.

Label the exterior tab and use the inside space to write information.

PROJECT FORMAT
Use 11" × 17" or 12" × 18" paper on the horizontal axis to make a large project book.

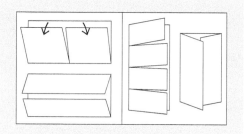

Variations
Paper can be folded horizontally, like a *hamburger* or vertically, like a *hot dog*.

A

B

C Half-books can be folded so that one side is ½ inch longer than the other side. A title or question can be written on the extended tab.

Worksheet Foldable or Folded Book® By Dinah Zike

Step 1 Make a half-book (see above) using work sheets, internet print-outs, diagrams, or maps.

Step 2 Fold it in half again.

Variations

A This folded sheet as a small book with two pages can be used for comparing and contrasting, cause and effect, or other skills.

B When the sheet of paper is open, the four sections can be used separately or used collectively to show sequences or steps.

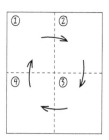

Two-Tab and Concept-Map Foldable® By Dinah Zike

Step 1 Fold a sheet of notebook or copy paper in half vertically or horizontally.

Step 2 Fold it in half again, as shown.

Step 3 Unfold once and cut along the fold line or valley of the top flap to make two flaps.

Variations

A Concept maps can be made by leaving a ½ inch tab at the top when folding the paper in half. Use arrows and labels to relate topics to the primary concept.

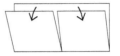

B Use two sheets of paper to make multiple page tab books. Glue or staple books together at the top fold.

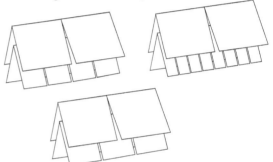

Three-Quarter Foldable® By Dinah Zike

Step 1 Make a two-tab book (see above) and cut the left tab off at the top of the fold line.

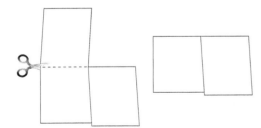

Variations

A Use this book to draw a diagram or a map on the exposed left tab. Write questions about the illustration on the top right tab and provide complete answers on the space under the tab.

B Compose a self-test using multiple choice answers for your questions. Include the correct answer with three wrong responses. The correct answers can be written on the back of the book or upside down on the bottom of the inside page.

Foldables Handbook • SR-31

Three-Tab Foldable® By Dinah Zike

Step 1 Fold a sheet of paper in half horizontally.

Step 2 Fold into thirds.

Step 3 Unfold and cut along the folds of the top flap to make three sections.

Variations

A Before cutting the three tabs draw a Venn diagram across the front of the book.

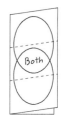

B Make a space to use for titles or concept maps by leaving a ½ inch tab at the top when folding the paper in half.

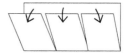

Four-Tab Foldable® By Dinah Zike

Step 1 Fold a sheet of paper in half horizontally.

Step 2 Fold in half and then fold each half as shown below.

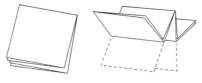

Step 3 Unfold and cut along the fold lines of the top flap to make four tabs.

Variations

A Make a space to use for titles or concept maps by leaving a ½ inch tab at the top when folding the paper in half.

B Use the book on the vertical axis, with or without an extended tab.

Folding Fifths for a Foldable® By Dinah Zike

Step 1 Fold a sheet of paper in half horizontally.

Step 2 Fold again so one-third of the paper is exposed and two-thirds are covered.

Step 3 Fold the two-thirds section in half.

Step 4 Fold the one-third section, a single thickness, backward to make a fold line.

Variations

A Unfold and cut along the fold lines to make five tabs.

B Make a five-tab book with a ½ inch tab at the top (see two-tab instructions).

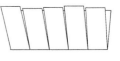

C Use 11" × 17" or 12" × 18" paper and fold into fifths for a five-column and/or row table or chart.

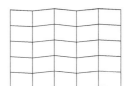

Folded Table or Chart, and Trifold Foldable® By Dinah Zike

Step 1 Fold a sheet of paper in the required number of vertical columns for the table or chart.

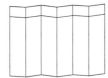

Step 2 Fold the horizontal rows needed for the table or chart.

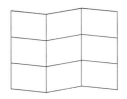

PROJECT FORMAT
Use 11" × 17" or 12" × 18" paper and fold it to make a large trifold project book or larger tables and charts.

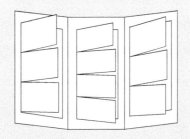

Variations

A Make a trifold by folding the paper into thirds vertically or horizontally.

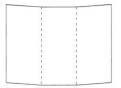

B Make a trifold book. Unfold it and draw a Venn diagram on the inside.

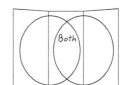

Two or Three-Pockets Foldable® By Dinah Zike

Step 1 Fold up the long side of a horizontal sheet of paper about 5 cm.

Step 2 Fold the paper in half.

Step 3 Open the paper and glue or staple the outer edges to make two compartments.

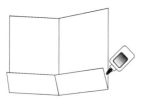

Variations

A Make a multi-page booklet by gluing several pocket books together.

B Make a three-pocket book by using a trifold (see previous instructions).

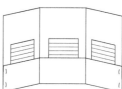

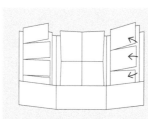

PROJECT FORMAT
Use 11" × 17" or 12" × 18" paper and fold it horizontally to make a large multi-pocket project.

- -

Matchbook Foldable® By Dinah Zike

Step 1 Fold a sheet of paper almost in half and make the back edge about 1–2 cm longer than the front edge.

Step 2 Find the midpoint of the shorter flap.

Step 3 Open the paper and cut the short side along the midpoint making two tabs.

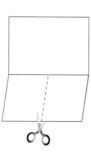

Step 4 Close the book and fold the tab over the short side.

Variations

A Make a single-tab matchbook by skipping Steps 2 and 3.

B Make two smaller matchbooks by cutting the single-tab matchbook in half.

Shutterfold Foldable® By Dinah Zike

Step 1 Begin as if you were folding a vertical sheet of paper in half, but instead of creasing the paper, pinch it to show the midpoint.

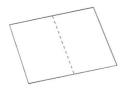

Step 2 Fold the top and bottom to the middle and crease the folds.

Variations

A Use the shutterfold on the horizontal axis.

B Create a center tab by leaving .5–2 cm between the flaps in Step 2.

PROJECT FORMAT
Use 11" × 17" or 12" × 18" paper and fold it to make a large shutterfold project.

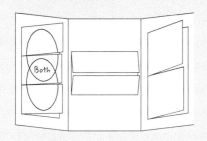

Four-Door Foldable® By Dinah Zike

Step 1 Make a shutterfold (see above).

Step 2 Fold the sheet of paper in half.

Step 3 Open the last fold and cut along the inside fold lines to make four tabs.

Variations

A Use the four-door book on the opposite axis.

B Create a center tab by leaving .5–2 cm between the flaps in Step 1.

Foldables Handbook • SR-35

Bound Book Foldable® By Dinah Zike

Step 1 Fold three sheets of paper in half. Place the papers in a stack, leaving about .5 cm between each top fold. Mark all three sheets about 3 cm from the outer edges.

Step 2 Using two of the sheets, cut from the outer edges to the marked spots on each side. On the other sheet, cut between the marked spots.

Step 3 Take the two sheets from Step 1 and slide them through the cut in the third sheet to make a 12-page book.

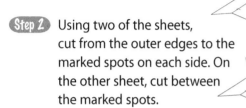

Step 4 Fold the bound pages in half to form a book.

Variation

A Use two sheets of paper to make an eight-page book, or increase the number of pages by using more than three sheets.

PROJECT FORMAT
Use two or more sheets of 11" × 17" or 12" × 18" paper and fold it to make a large bound book project.

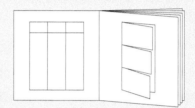

Accordian Foldable® By Dinah Zike

Step 1 Fold the selected paper in half vertically, like a *hamburger*.

Step 2 Cut each sheet of folded paper in half along the fold lines.

Step 3 Fold each half-sheet almost in half, leaving a 2 cm tab at the top.

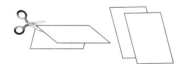

Step 4 Fold the top tab over the short side, then fold it in the opposite direction.

Variations

A Glue the straight edge of one paper inside the tab of another sheet. Leave a tab at the end of the book to add more pages.

B Tape the straight edge of one paper to the tab of another sheet, or just tape the straight edges of nonfolded paper end to end to make an accordian.

C Use whole sheets of paper to make a large accordian.

Layered Foldable® By Dinah Zike

Step 1 Stack two sheets of paper about 1–2 cm apart. Keep the right and left edges even.

Step 2 Fold up the bottom edges to form four tabs. Crease the fold to hold the tabs in place.

Step 3 Staple along the folded edge, or open and glue the papers together at the fold line.

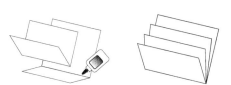

Variations

A Rotate the book so the fold is at the top or to the side.

B Extend the book by using more than two sheets of paper.

. . . .

Envelope Foldable® By Dinah Zike

Step 1 Fold a sheet of paper into a *taco*. Cut off the tab at the top.

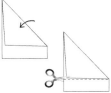

Step 2 Open the *taco* and fold it the opposite way making another *taco* and an X-fold pattern on the sheet of paper.

Step 3 Cut a map, illustration, or diagram to fit the inside of the envelope.

Step 4 Use the outside tabs for labels and inside tabs for writing information.

Variations

A Use 11" × 17" or 12" × 18" paper to make a large envelope.

B Cut off the points of the four tabs to make a window in the middle of the book.

Sentence Strip Foldable® By Dinah Zike

Step 1 Fold two sheets of paper in half vertically, like a *hamburger*.

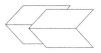

Step 2 Unfold and cut along fold lines making four half sheets.

Step 3 Fold each half sheet in half horizontally, like a *hot dog*.

Step 4 Stack folded horizontal sheets evenly and staple together on the left side.

Step 5 Open the top flap of the first sentence strip and make a cut about 2 cm from the stapled edge to the fold line. This forms a flap that can be raised and lowered. Repeat this step for each sentence strip.

Variations

A Expand this book by using more than two sheets of paper.

B Use whole sheets of paper to make large books.

Pyramid Foldable® By Dinah Zike

Step 1 Fold a sheet of paper into a *taco*. Crease the fold line, but do not cut it off.

Step 2 Open the folded sheet and refold it like a *taco* in the opposite direction to create an X-fold pattern.

Step 3 Cut one fold line as shown, stopping at the center of the X-fold to make a flap.

Step 4 Outline the fold lines of the X-fold. Label the three front sections and use the inside spaces for notes. Use the tab for the title.

Step 5 Glue the tab into a project book or notebook. Use the space under the pyramid for other information.

Step 6 To display the pyramid, fold the flap under and secure with a paper clip, if needed.

Single-Pocket or One-Pocket Foldable® By Dinah Zike

Step 1 Using a large piece of paper on a vertical axis, fold the bottom edge of the paper upwards, about 5 cm.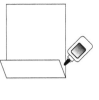

Step 2 Glue or staple the outer edges to make a large pocket.

PROJECT FORMAT
Use 11" × 17" or 12" × 18" paper and fold it vertically or horizontally to make a large pocket project.

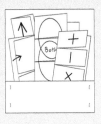

Variations

A Make the one-pocket project using the paper on the horizontal axis.

B To store materials securely inside, fold the top of the paper almost to the center, leaving about 2–4 cm between the paper edges. Slip the Foldables through the opening and under the top and bottom pockets.

Multi-Tab Foldable® By Dinah Zike

Step 1 Fold a sheet of notebook paper in half like a *hot dog*.

Step 2 Open the paper and on one side cut every third line. This makes ten tabs on wide ruled notebook paper and twelve tabs on college ruled.

Step 3 Label the tabs on the front side and use the inside space for definitions or other information.

Variation

A Make a tab for a title by folding the paper so the holes remain uncovered. This allows the notebook Foldable to be stored in a three-hole binder.

Reference Handbook

PERIODIC TABLE OF THE ELEMENTS

Element — Hydrogen
Atomic number — 1
Symbol — H
Atomic mass — 1.01
State of matter

- Gas
- Liquid
- Solid
- Synthetic

A column in the periodic table is called a **group**.

A row in the periodic table is called a **period**.

The number in parentheses is the mass number of the longest lived isotope for that element.

Period	1	2	3	4	5	6	7	8	9
1	Hydrogen 1 H 1.01								
2	Lithium 3 Li 6.94	Beryllium 4 Be 9.01							
3	Sodium 11 Na 22.99	Magnesium 12 Mg 24.31							
4	Potassium 19 K 39.10	Calcium 20 Ca 40.08	Scandium 21 Sc 44.96	Titanium 22 Ti 47.87	Vanadium 23 V 50.94	Chromium 24 Cr 52.00	Manganese 25 Mn 54.94	Iron 26 Fe 55.85	Cobalt 27 Co 58.93
5	Rubidium 37 Rb 85.47	Strontium 38 Sr 87.62	Yttrium 39 Y 88.91	Zirconium 40 Zr 91.22	Niobium 41 Nb 92.91	Molybdenum 42 Mo 95.96	Technetium 43 Tc (98)	Ruthenium 44 Ru 101.07	Rhodium 45 Rh 102.91
6	Cesium 55 Cs 132.91	Barium 56 Ba 137.33	Lanthanum 57 La 138.91	Hafnium 72 Hf 178.49	Tantalum 73 Ta 180.95	Tungsten 74 W 183.84	Rhenium 75 Re 186.21	Osmium 76 Os 190.23	Iridium 77 Ir 192.22
7	Francium 87 Fr (223)	Radium 88 Ra (226)	Actinium 89 Ac (227)	Rutherfordium 104 Rf (267)	Dubnium 105 Db (268)	Seaborgium 106 Sg (271)	Bohrium 107 Bh (272)	Hassium 108 Hs (270)	Meitnerium 109 Mt (276)

Lanthanide series

Cerium 58 Ce 140.12	Praseodymium 59 Pr 140.91	Neodymium 60 Nd 144.24	Promethium 61 Pm (145)	Samarium 62 Sm 150.36	Europium 63 Eu 151.96

Actinide series

Thorium 90 Th 232.04	Protactinium 91 Pa 231.04	Uranium 92 U 238.03	Neptunium 93 Np (237)	Plutonium 94 Pu (244)	Americium 95 Am (243)

Legend

- Metal
- Metalloid
- Nonmetal
- Recently discovered

Periodic Table (partial)

Group 18

13	14	15	16	17	18
					Helium 2 He 4.00
Boron 5 B 10.81	Carbon 6 C 12.01	Nitrogen 7 N 14.01	Oxygen 8 O 16.00	Fluorine 9 F 19.00	Neon 10 Ne 20.18
Aluminum 13 Al 26.98	Silicon 14 Si 28.09	Phosphorus 15 P 30.97	Sulfur 16 S 32.07	Chlorine 17 Cl 35.45	Argon 18 Ar 39.95

10	11	12	13	14	15	16	17	18
Nickel 28 Ni 58.69	Copper 29 Cu 63.55	Zinc 30 Zn 65.38	Gallium 31 Ga 69.72	Germanium 32 Ge 72.64	Arsenic 33 As 74.92	Selenium 34 Se 78.96	Bromine 35 Br 79.90	Krypton 36 Kr 83.80
Palladium 46 Pd 106.42	Silver 47 Ag 107.87	Cadmium 48 Cd 112.41	Indium 49 In 114.82	Tin 50 Sn 118.71	Antimony 51 Sb 121.76	Tellurium 52 Te 127.60	Iodine 53 I 126.90	Xenon 54 Xe 131.29
Platinum 78 Pt 195.08	Gold 79 Au 196.97	Mercury 80 Hg 200.59	Thallium 81 Tl 204.38	Lead 82 Pb 207.20	Bismuth 83 Bi 208.98	Polonium 84 Po (209)	Astatine 85 At (210)	Radon 86 Rn (222)
Darmstadtium 110 Ds (281)	Roentgenium 111 Rg (280)	Copernicium 112 Cn (285)	* Ununtrium 113 Uut (284)	* Ununquadium 114 Uuq (289)	* Ununpentium 115 Uup (288)	* Ununhexium 116 Uuh (293)		* Ununoctium 118 Uuo (294)

* The names and symbols for elements 113–116 and 118 are temporary. Final names will be selected when the elements' discoveries are verified.

Gadolinium 64 Gd 157.25	Terbium 65 Tb 158.93	Dysprosium 66 Dy 162.50	Holmium 67 Ho 164.93	Erbium 68 Er 167.26	Thulium 69 Tm 168.93	Ytterbium 70 Yb 173.05	Lutetium 71 Lu 174.97
Curium 96 Cm (247)	Berkelium 97 Bk (247)	Californium 98 Cf (251)	Einsteinium 99 Es (252)	Fermium 100 Fm (257)	Mendelevium 101 Md (258)	Nobelium 102 No (259)	Lawrencium 103 Lr (262)

Reference Handbook • **SR-41**

Topographic Map Symbols

Topographic Map Symbols

Symbol	Description	Symbol	Description
	Primary highway, hard surface		Index contour
	Secondary highway, hard surface		Supplementary contour
	Light-duty road, hard or improved surface		Intermediate contour
	Unimproved road		Depression contours
	Railroad: single track		
	Railroad: multiple track		Boundaries: national
	Railroads in juxtaposition		State
			County, parish, municipal
	Buildings		Civil township, precinct, town, barrio
	Schools, church, and cemetery		Incorporated city, village, town, hamlet
	Buildings (barn, warehouse, etc.)		Reservation, national or state
	Wells other than water (labeled as to type)		Small park, cemetery, airport, etc.
	Tanks: oil, water, etc. (labeled only if water)		Land grant
	Located or landmark object; windmill		Township or range line, U.S. land survey
	Open pit, mine, or quarry; prospect		Township or range line, approximate location
	Marsh (swamp)		
	Wooded marsh		Perennial streams
	Woods or brushwood		Elevated aqueduct
	Vineyard		Water well and spring
	Land subject to controlled inundation		Small rapids
	Submerged marsh		Large rapids
	Mangrove		Intermittent lake
	Orchard		Intermittent stream
	Scrub		Aqueduct tunnel
	Urban area		Glacier
			Small falls
x7369	Spot elevation		Large falls
670	Water elevation		Dry lake bed

Rocks

Rocks

Rock Type	Rock Name	Characteristics
Igneous (intrusive)	Granite	Large mineral grains of quartz, feldspar, hornblende, and mica. Usually light in color.
	Diorite	Large mineral grains of feldspar, hornblende, and mica. Less quartz than granite. Intermediate in color.
	Gabbro	Large mineral grains of feldspar, augite, and olivine. No quartz. Dark in color.
Igneous (extrusive)	Rhyolite	Small mineral grains of quartz, feldspar, hornblende, and mica, or no visible grains. Light in color.
	Andesite	Small mineral grains of feldspar, hornblende, and mica or no visible grains. Intermediate in color.
	Basalt	Small mineral grains of feldspar, augite, and possibly olivine or no visible grains. No quartz. Dark in color.
	Obsidian	Glassy texture. No visible grains. Volcanic glass. Fracture looks like broken glass.
	Pumice	Frothy texture. Floats in water. Usually light in color.
Sedimentary (detrital)	Conglomerate	Coarse grained. Gravel or pebble-size grains.
	Sandstone	Sand-sized grains 1/16 to 2 mm.
	Siltstone	Grains are smaller than sand but larger than clay.
	Shale	Smallest grains. Often dark in color. Usually platy.
Sedimentary (chemical or organic)	Limestone	Major mineral is calcite. Usually forms in oceans and lakes. Often contains fossils.
	Coal	Forms in swampy areas. Compacted layers of organic material, mainly plant remains.
Sedimentary (chemical)	Rock Salt	Commonly forms by the evaporation of seawater.
Metamorphic (foliated)	Gneiss	Banding due to alternate layers of different minerals, of different colors. Parent rock often is granite.
	Schist	Parallel arrangement of sheetlike minerals, mainly micas. Forms from different parent rocks.
	Phyllite	Shiny or silky appearance. May look wrinkled. Common parent rocks are shale and slate.
	Slate	Harder, denser, and shinier than shale. Common parent rock is shale.
Metamorphic (nonfoliated)	Marble	Calcite or dolomite. Common parent rock is limestone.
	Soapstone	Mainly of talc. Soft with greasy feel.
	Quartzite	Hard with interlocking quartz crystals. Common parent rock is sandstone.

Minerals

Minerals

Mineral (formula)	Color	Streak	Hardness Pattern	Breakage Properties	Uses and Other
Graphite (C)	black to gray	black to gray	1–1.5	basal cleavage (scales)	pencil lead, lubricants for locks, rods to control some small nuclear reactions, battery poles
Galena (PbS)	gray	gray to black	2.5	cubic cleavage perfect	source of lead, used for pipes, shields for X rays, fishing equipment sinkers
Hematite (Fe_2O_3)	black or reddish-brown	reddish-brown	5.5–6.5	irregular fracture	source of iron; converted to pig iron, made into steel
Magnetite (Fe_3O_4)	black	black	6	conchoidal fracture	source of iron, attracts a magnet
Pyrite (FeS_2)	light, brassy, yellow	greenish-black	6–6.5	uneven fracture	fool's gold
Talc ($Mg_3Si_4O_{10}(OH)_2$)	white, greenish	white	1	cleavage in one direction	used for talcum powder, sculptures, paper, and tabletops
Gypsum ($CaSO_4 \cdot 2H_2O$)	colorless, gray, white, brown	white	2	basal cleavage	used in plaster of paris and dry wall for building construction
Sphalerite (ZnS)	brown, reddish-brown, greenish	light to dark brown	3.5–4	cleavage in six directions	main ore of zinc; used in paints, dyes, and medicine
Muscovite ($KAl_3Si_3O_{10}(OH)_2$)	white, light gray, yellow, rose, green	colorless	2–2.5	basal cleavage	occurs in large, flexible plates; used as an insulator in electrical equipment, lubricant
Biotite ($K(Mg,Fe)_3(AlSi_3O_{10})(OH)_2$)	black to dark brown	colorless	2.5–3	basal cleavage	occurs in large, flexible plates
Halite (NaCl)	colorless, red, white, blue	colorless	2.5	cubic cleavage	salt; soluble in water; a preservative

Minerals

Minerals

Mineral (formula)	Color	Streak	Hardness	Breakage Pattern	Uses and Other Properties
Calcite ($CaCO_3$)	colorless, white, pale blue	colorless, white	3	cleavage in three directions	fizzes when HCl is added; used in cements and other building materials
Dolomite ($CaMg(CO_3)_2$)	colorless, white, pink, green, gray, black	white	3.5–4	cleavage in three directions	concrete and cement; used as an ornamental building stone
Fluorite (CaF_2)	colorless, white, blue, green, red, yellow, purple	colorless	4	cleavage in four directions	used in the manufacture of optical equipment; glows under ultraviolet light
Hornblende $(CaNa)_{2-3}(Mg,Al,Fe)_5-(Al,Si)_2 Si_6O_{22}(OH)_2$	green to black	gray to white	5–6	cleavage in two directions	will transmit light on thin edges; 6-sided cross section
Feldspar ($KAlSi_3O_8$) ($NaAlSi_3O_8$), ($CaAl_2Si_2O_8$)	colorless, white to gray, green	colorless	6	two cleavage planes meet at 90° angle	used in the manufacture of ceramics
Augite $((Ca,Na)(Mg,Fe,Al)(Al,Si)_2O_6)$	black	colorless	6	cleavage in two directions	square or 8-sided cross section
Olivine $((Mg,Fe)_2SiO_4)$	olive, green	none	6.5–7	conchoidal fracture	gemstones, refractory sand
Quartz (SiO_2)	colorless, various colors	none	7	conchoidal fracture	used in glass manufacture, electronic equipment, radios, computers, watches, gemstones

Weather Map Symbols

Sample Station Model

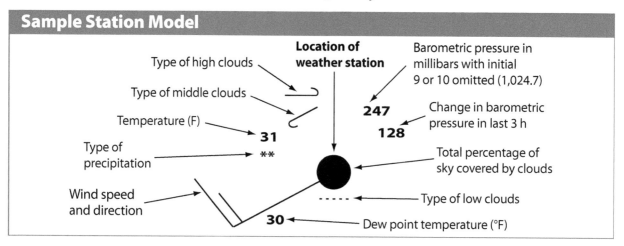

Sample Plotted Report at Each Station

Precipitation		Wind Speed and Direction		Sky Coverage		Some Types of High Clouds	
≡	Fog	○	0 calm	○	No cover	⌒	Scattered cirrus
★	Snow	/	1–2 knots	⊙	1/10 or less	⌒⌒	Dense cirrus in patches
●	Rain	⌐	3–7 knots	◔	2/10 to 3/10	⌒⌒⌒	Veil of cirrus covering entire sky
⊼	Thunderstorm	⌐⌐	8–12 knots	◔	4/10	⌒⌒	Cirrus not covering entire sky
,	Drizzle	⌐⌐⌐	13–17 knots	◐	–		
▽	Showers	⌐⌐⌐⌐	18–22 knots	◕	6/10		
		⌐⌐⌐⌐⌐	23–27 knots	◕	7/10		
		◢	48–52 knots	◕	Overcast with openings		
		1 knot = 1.852 km/h		●	Completely overcast		

Some Types of Middle Clouds		Some Types of Low Clouds		Fronts and Pressure Systems	
∠	Thin altostratus layer	⌒	Cumulus of fair weather	Ⓗ or High Ⓛ or Low	Center of high- or low-pressure system
∠∠	Thick altostratus layer	∪	Stratocumulus	▲▲▲▲	Cold front
⌒	Thin altostratus in patches	-----	Fractocumulus of bad weather	●●●●	Warm front
⌒⌒	Thin altostratus in bands	—	Stratus of fair weather	▲●▲●	Occluded front
				▲●▲●	Stationary front

Glossary/Glosario

Multilingual eGlossary

A science multilingual glossary is available on the science Web site. The glossary includes the following languages.

Arabic	Hmong	Tagalog
Bengali	Korean	Urdu
Chinese	Portuguese	Vietnamese
English	Russian	
Haitian Creole	Spanish	

Cómo usar el glosario en español:
1. Busca el término en inglés que desees encontrar.
2. El término en español, junto con la definición, se encuentran en la columna de la derecha.

Pronunciation Key

Use the following key to help you sound out words in the glossary.

a	back (BAK)		ew	food (FEWD)
ay	day (DAY)		yoo	pure (PYOOR)
ah	father (FAH thur)		yew	few (FYEW)
ow	flower (FLOW ur)		uh	comma (CAH muh)
ar	car (CAR)		u (+ con)	rub (RUB)
e	less (LES)		sh	shelf (SHELF)
ee	leaf (LEEF)		ch	nature (NAY chur)
ih	trip (TRIHP)		g	gift (GIHFT)
i (i + com + e)	idea (i DEE uh)		j	gem (JEM)
oh	go (GOH)		ing	sing (SING)
aw	soft (SAWFT)		zh	vision (VIH zhun)
or	orbit (OR buht)		k	cake (KAYK)
oy	coin (COYN)		s	seed, cent (SEED, SENT)
oo	foot (FOOT)		z	zone, raise (ZOHN, RAYZ)

English / Español — A

apparent magnitude/Big Bang theory — magnitud aparente/Teoría del Big Bang

apparent magnitude: a measure of how bright an object appears from Earth. (p. 805)

asteroid: a small, rocky object that orbits the Sun. (p. 763)

astrobiology: the study of the origin, development, distribution, and future of life on Earth and in the universe. (p. 711)

astronomical unit (AU): the average distance from Earth to the Sun—about 150 million km. (pp. 764, 804)

magnitud aparente: medida del brillo de un objeto visto desde la Tierra. (pág. 805)

asteroide: objeto pequeño y rocoso que orbita el Sol. (pág. 763)

astrobiología: estudio del origen, desarrollo, distribución y futuro de la vida en la Tierra y en el universo. (pág. 711)

unidad astronómica (UA): distancia media entre la Tierra y el Sol, aproximadamente 150 millones de km. (páges. 764, 804)

B

Big Bang theory: the scientific theory that states that the universe began from one point and has been expanding and cooling ever since. (p. 830)

Teoría del Big Bang: teoría científica que establece que el universo se originó de un punto y se ha ido expandiendo y enfriando desde entonces. (pág. 830)

black hole/greenhouse effect

black hole: an object whose gravity is so great that no light can escape. (p. 820)

C

chromosphere: the orange-red layer above the photosphere of a star. (p. 810)

comet: a small, rocky, icy object that orbits the Sun. (p. 763)

convection zone: layer of a star where hot gas moves up toward the surface and cooler gas moves deeper into the interior. (p. 810)

corona: the wide, outermost layer of a star's atmosphere. (p. 810)

D

dark matter: matter that emits no light at any wavelength. (p. 825)

Doppler shift: the shift to a different wavelength on the electromagnetic spectrum. (p. 830)

E

electromagnetic (ih lek troh mag NEH tik) spectrum: the entire range of radiant energy carried by electromagnetic waves. (p. 690)

equinox: when Earth's rotation axis is tilted neither toward nor away from the Sun. (p. 731)

extraterrestrial (ek struh tuh RES tree ul) life: life that originates outside Earth. (p. 711)

G

galaxy: a huge collection of stars, gas, and dust. (p. 825)

Galilean moons: the four largest of Jupiter's 63 moons; discovered by Galileo. (p. 779)

greenhouse effect: the natural process that occurs when certain gases in the atmosphere absorb and reradiate thermal energy from the Sun. (p. 771)

agujero negro/efecto invernadero

agujero negro: objeto cuya gravedad es tan grande que la luz no puede escapar. (pág. 820)

cromosfera: capa de color rojo anaranjado arriba de la fotosfera de una estrella. (pág. 810)

cometa: objeto pequeño, rocoso y helado que orbita el Sol. (pág. 763)

zona de convección: capa de una estrella donde el gas caliente se mueve hacia arriba de la superficie y el gas más frío se mueve más profundo hacia el interior. (pág. 810)

corona: capa extensa más externa de la atmósfera de una estrella. (pág. 810)

materia oscura: materia que no emite luz a ninguna longitud de onda. (pág. 825)

efecto Doppler: cambio a una longitud de onda diferente en el espectro electromagnético. (pág. 830)

espectro electromagnético: gama completa de energía radiante transportada por las ondas electromagnéticas. (pág. 690)

equinoccio: cuando el eje de rotación de la Tierra se inclina sin acercarse ni alejarse del Sol. (pág. 731)

vida extraterrestre: vida que se origina fuera de la Tierra. (pág. 711)

galaxia: conjunto enorme de estrellas, gas, y polvo. (pág. 825)

lunas de Galileo: las cuatro lunas más grandes de las 63 lunas de Júpiter; descubiertas por Galileo. (pág. 779)

efecto invernadero: proceso natural que ocurre cuando ciertos gases en la atmósfera absorben y vuelven a irradiar la energía térmica del Sol. (pág. 771)

Hertzsprung-Russell diagram/nuclear fusion   **diagrama de Hertzsprung-Russell/fusión nuclear**

H

Hertzsprung-Russell diagram: a graph that plots luminosity v. temperature of stars. (p. 813)

diagrama de Hertzsprung-Russell: diagrama que traza la luminosidad frente a la temperatura de las estrellas. (pág. 813)

I

impact crater: a round depression formed on the surface of a planet, moon, or other space object by the impact of a meteorite. (p. 788)

cráter de impacto: depresión redonda formada en la superficie de un planeta, luna u otro objeto espacial debido al impacto de un meteorito. (pág. 788)

L

light-year: the distance light travels in one year. (p. 804)

luminosity (lew muh NAH sih tee): the true brightness of an object. (p. 125)

lunar eclipse: an occurrence during which the Moon moves into Earth's shadow. (p. 746)

lunar: term that refers to anything related to the Moon. (p. 701)

año luz: distancia que recorre la luz en un año. (pág. 804)

luminosidad: brillantez real de un objeto. (pág. 125)

eclipse lunar: ocurrencia durante la cual la Luna se mueve hacia la zona de sombra de la Tierra. (pág. 746)

lunar: término que hace referencia a todo lo relacionado con la luna. (pág. 701)

M

maria (MAR ee uh): the large, dark, flat areas on the Moon. (p. 736)

meteor: a meteoroid that has entered Earth's atmosphere and produces a streak of light. (p. 788)

meteorite: a meteoroid that strikes a planet or a moon. (p. 788)

meteoroid: a small, rocky particle that moves through space. (p. 788)

mares: áreas extensas, oscuras y planas en la Luna. (pág. 736)

meteoro: meteorito que ha entrado a la atmósfera de la Tierra y produce un haz de luz. (pág. 788)

meteorito: meteoroide que impacta un planeta o una luna. (pág. 788)

meteoroide: partícula rocosa pequeña que se mueve por el espacio. (pág. 788)

N

nebula: a cloud of gas and dust. (p. 817)

neutron star: a dense core of neutrons that remains after a supernova. (p. 820)

nuclear fusion: a process that occurs when the nuclei of several atoms combine into one larger nucleus. (p. 809)

nebulosa: nube de gas y polvo. (pág. 817)

estrella de neutrones: núcleo denso de neutrones que queda después de una supernova. (pág. 820)

fusión nuclear: proceso que ocurre cuando los núcleos de varios átomos se combinan en un núcleo mayor. (pág. 809)

orbit/rotation   **órbita/rotación**

O

orbit: the path an object follows as it moves around another object. (p. 726)

órbita: trayectoria que un objeto sigue a medida que se mueve alrededor de otro objeto. (pág. 726)

P

penumbra: the lighter part of a shadow where light is partially blocked. (p. 743)

period of revolution: the time it takes an object to travel once around the Sun. (p. 764)

period of rotation: the time it takes an object to complete one rotation. (p. 764)

phase: the lit part of the Moon or a planet that can be seen from Earth. (p. 738)

photosphere: the apparent surface of a star. (p. 810)

Project Apollo: a series of space missions designed to send people to the Moon. (p. 702)

penumbra: parte más clara de una sombra donde la luz se bloquea parcialmente. (pág. 743)

período de revolución: tiempo que gasta un objeto en dar una vuelta alrededor del Sol. (pág. 764)

período de rotación: tiempo que gasta un objeto para completar una rotación. (pág. 764)

fase: parte iluminada de la Luna o de un planeta que se ve desde la Tierra. (pág. 738)

fotosfera: superficie luminosa de una estrella. (pág. 810)

Proyecto Apolo: serie de misiones espaciales diseñadas para enviar personas a la Luna. (pág. 702)

R

radiative zone: a shell of cooler hydrogen above a star's core. (p. 810)

radio telescope: a telescope that collects radio waves and some microwaves using an antenna that looks like a TV satellite dish. (p. 693)

reflecting telescope: a telescope that uses a curved mirror to concentrate light from distant objects. (p. 692)

refracting telescope: a telescope that uses a convex lens to concentrate light from distant objects. (p. 692)

revolution: the orbit of one object around another object. (p. 726)

rocket: a vehicle propelled by the exhaust made from burning fuel. (p. 699)

rotation axis: the line on which an object rotates. (p. 727)

rotation: the spin of an object around its axis. (p. 727)

zona radiativa: capa de hidrógeno más frío por encima del núcleo de una estrella. (pág. 810)

radiotelescopio: telescopio que recoge ondas de radio y algunas microondas por medio de una antena parecida a una antena parabólica de TV. (pág. 693)

telescopio reflector: telescopio que tiene un espejo para reunir y enfocar luz de objetos lejanos. (pág. 692)

telescopio refractor: telescopio que usa un lente convexo para enfocar la luz de objetos lejanos. (pág. 692)

revolución: movimiento de un objeto alrededor de otro objeto. (pág. 726)

cohete: vehículo propulsado por gases de escape producidos por la ignición de combustible. (pág. 699)

eje de rotación: línea sobre la cual un objeto rota. (pág. 727)

rotación: movimiento giratorio de un objeto sobre su eje. (pág. 727)

satellite/waning phases

satélite/fases menguantes

S

satellite: any small object that orbits a larger object other than a star. (p. 700)

solar eclipse: an occurrence during which the Moon's shadow appears on Earth's surface. (p. 744)

solstice: when Earth's rotation axis is tilted directly toward or away from the Sun. (p. 731)

space probe: an uncrewed spacecraft sent from Earth to explore objects in space. (p. 701)

space shuttles: reusable spacecraft that transport people and materials to and from space. (p. 702)

spectroscope: an instrument that spreads light into different wavelengths. (p. 803)

star: a large sphere of hydrogen gas, held together by gravity, that is hot enough for nuclear reactions to occur in its core. (p. 809)

supernova: an enormous explosion that destroys a star. (p. 819)

satélite: cualquier objeto pequeño que orbita un objeto más grande diferente de una estrella. (pág. 700)

eclipse solar: acontecimiento durante el cual la sombra de la Luna aparece sobre la superficie de la Tierra. (pág. 744)

solsticio: cuando el eje de rotación de la Tierra se inclina acercándose o alejándose del Sol. (pág. 731)

sonda espacial: nave espacial sin tripulación enviada desde la Tierra para explorar objetos en el espacio. (pág. 701)

transbordador espacial: nave espacial reutilizable que transporta personas y materiales hacia y desde el espacio. (pág. 702)

espectroscopio: instrumento utilizado para propagar la luz en diferentes longitudes de onda. (pág. 803)

estrella: esfera enorme de gas de hidrógeno, que se mantiene unida por la gravedad, lo suficientemente caliente para producir reacciones nucleares en el núcleo. (pág. 809)

supernova: explosión enorme que destruye una estrella. (pág. 819)

T

terrestrial planets: Earth and the other inner planets that are closest to the Sun including Mercury, Venus, and Mars. (p. 769)

tide: the periodic rise and fall of the ocean's surface caused by the gravitational force between Earth and the Moon, and Earth and the Sun. (p. 747)

planetas terrestres: la Tierra y otros planetas interiores que están más cerca del Sol, incluidos Mercurio, Venus y Marte. (pág. 769)

marea: ascenso y descenso periódico de la superficie del océano causados por la fuerza gravitacional entre la Tierra y la Luna, y entre la Tierra y el Sol. (pág. 747)

U

umbra: the central, darker part of a shadow where light is totally blocked. (p. 743)

umbra: parte central más oscura de una sombra donde la luz está completamente bloqueada. (pág. 743)

W

waning phases: phases of the Moon during which less of the Moon's near side is lit each night. (p. 738)

fases menguantes: fases de la Luna durante las cuales el lado cercano de la Luna está menos iluminado cada noche. (pág. 738)

waxing phases: phases of the Moon during which more of the Moon's near side is lit each night. (p. 738)

white dwarf: a hot, dense, slowly cooling sphere of carbon. (p. 819)

fases crecientes: fases de la Luna durante las cuales el lado cercano de la Luna está más iluminado cada noche. (pág. 738)

enana blanca: esfera de carbón caliente y densa que se enfría lentamente. (pág. 819)

Index

Academic Vocabulary   *Italic numbers* = illustration/photo   **Bold numbers** = vocabulary term   **Graph(s)**
lab = indicates entry is used in a lab on this page

A

Academic Vocabulary, 710, 728, 805. *See also* **Vocabulary**
Accretion hypothesis, 767
Adaptive optics, 693, *693*
Aldrin, Buzz, 702
Apollo 697, 702, *702*
Apollo Space Program, 741
Apparent magnitude, 805
Armstrong, Neil, 702, 741
Asteroid(s)
    explanation of, *763*, **763,** 786
    formation of, 767, 785 *lab*
Astrobiology
    explanation of, **711**
Astronaut(s), 741
    early background of, 702, *702*
Astronomical unit (AU), 764, **804,** *804*
Astronomy
    naked-eye, 802
Astrophysicist(s), 783
Atmosphere
    of Earth, 772
Autumn. *See* **Fall**

B

Big Bang theory, 830
    Review, 837
Big Dipper, 807
Big Idea, 686, 716, 722, 752, 758, 792, 798, 834
    Review, 719, 755, 795, 837
Black hole(s), 820

C

Callisto, 779
Carbon dioxide
    on Venus, 771
Careers in Science, 767, 783
Cassini, 709, *709*
Cassini **orbiter,** 780
Cassiopeia A, 695, *695*
Ceres
    discovery of, 785
    as dwarf planet, 785, 786, *786*
    explanation of, *763*, **763,** 783, 785
Chapter Review, 718, 754–755, 794–795, 836–837
Chondrite meteorite(s)
    explanation of, 767
Chondrules, 767
Chromosphere, 810
Comets
    explanation of, **763,** 787, *787*
    formation of, 767
    short-period and long-period, 788
    structure of, 787
Common Use. *See* **Science Use v. Common Use**
Constellation(s)
    explanation of, 802
    use of scientific illustrations to locate, 807
Convection zone, 810
Corona, 810
Coronal mass ejections (CMEs), *811*, 815
Crater(s)
    impact, **788,** 788 *lab*
    on Moon, 736, *737*
Critical thinking, 696, 704, 713, 732, 740, 749, 766, 774, 782, 789, 795, 806, 814, 822, 831
Crust
    of Earth, 772

D

Dark energy, 830
Dark matter, 825
December solstice, 730, **731,** *731*
Distance
    units to measure, 3 *lab*, 764
Doppler shift, *830*, **830**
Dwarf planet(s)
    explanation of, 709, 763
    Pluto and, 783

E

Earth
    atmosphere on, 772
    composition of, 763
    data on, *772*
    distance between Sun and, 726
    movement of, 801
    required conditions for life on, 710 *lab*
    rotation around Sun, *726*, **726,** 726 *lab*, 727, *727*
    rotation of Moon around, 737, *737*, 738
    satellites that orbit, 712, *712*
    seasons on, 729, *729*, 730, 731, 733
    size of, *769*
    in solar system, 708
    structure of, 772, *773*
    temperature on surface of, 725 *lab*, 728
    tilt in rotational axis of, 727, *727*, 729, 733
Ebel, Denton, 767
Eclipse(s)
    lunar, 746–747, *746–747*
    solar, 744–745, *744*, *745*
Electromagnetic spectrum, *690*, **690,** 803, *803*
Electromagnetic wave(s)
    explanation of, 690
    from space, 694
    telescopes and, 692, 694, *694*
Ellipse, 765
Elliptical galaxy(ies), *826*
Elliptical orbit
    explanation of, 765
    modeling of, 765 *lab*
Energy
    radiant, 690
    from Sun, *726*, **726,** *728*, **728**
Equator, 728
Equinox
    explanation of, **731**
    March, 730, **731,** *731*
    September, 730, **731,** *731*
Eris, 763
    as dwarf planet, 785, 786, *786*
Europa, 711, *711*, 779
Explorer I, *700*, 701
Extraterrestrial life, **711**

F

Fall
    in Northern Hemisphere, 729, *729*
Flare(s), *811*
Flyby(s), *701*, 708
Foci, 765
Foldables, 688, 690, 700, 717, 731, 735, 746, 753, 764, 770, 778, 786, 793, 802, 809, 817, 827, 835

G

Galaxy(ies), **825**
    formation of, 821
    gravity in, 825
    groups of, 827
    identification of, 827 *lab*
    Milky Way as, 827, *828–829*
    types of, 826, *826*
Galilean moon(s), 778
Galileo Galilei, 779
Ganymede, 779
Gas giants. *See also* **Outer planets;** *specific planets*
    explanation of, 777
Gaspra, 763
Geologist(s), 767
Global Positioning System (GPS), 700
Globular cluster(s), *812*, **812**
Goddard, Robert, 700
Graph(s)
    function of, 823

I-2 • Index

Gravitational force
  causing objects to orbit the Sun, 762, *762*
  on Earth, 726, *726*
  of outer planets, 777
Gravity
  in black holes, 820
  effect on stars and planets, 821
  spacecraft travel and, 707 *lab*
Great Red Spot, 778
Greenhouse effect
  on Earth, 772, *772*
  explanation of, **771**
  on Venus, 771

H

Haumea
  as dwarf planet, 785, 786
Helium
  main-sequence stars and, 818, *818*
  as planetary nebulae, 821, *821*
Hertzsprung-Russell diagram, *813,* **813,** 818, 823
Highland(s)
  of Moon, 736, *737*
Hipparchus, 805
How it Works, 705, 815
Hubble Space Telescope, 694, *694,* 695, 827 *lab*
***Huygens* probe,** 780
Hydrogen
  main-sequence stars and, 818, *818*
  as planetary nebulae, 821, *821*

I

Impact crater(s)
  explanation of, **788**
  formation of, 788 *lab*
Inner planet(s). *See also* **Planet(s);** *specific planets*
  composition of, 769
  Earth, 772, *772*
  explanation of, *762,* 763
  Mars, 773, *773*
  Mercury, 770, *770*
  modeling of, 772 *lab*
  temperature on, 769 *lab*
  Venus, 771, *771*
International Astronomical Union (IAU), 785
International Space Station, 702, *702*
Interpret Graphics, 696, 704, 713, 732, 740, 749, 766, 774, 782, 789, 806, 814, 822
Io, 779
Irregular galaxy(ies), 826

J

James Webb Space Telescope, 695, *695*
June solstice, *730,* 731, *731*
Jupiter
  atmosphere of, 778
  data on, *778*
  explanation of, 709, 763, 778
  moons of, 779
  structure of, 778

K

Keck telescope(s), 692, *692*
Kepler telescope, 712, *712*
Key Concepts, 688, 698, 724, 734, 742, 760, 768, 776, 784, 800
  Check, 692, 694, 699, 700, 701, 703, 707, 710, 711, 712, 726, 728, 729, 737, 738, 744, 746, 748, 763, 764, 765, 770, 771, 773, 777, 778, 779, 780, 781, 785, 787, 788, 802, 803, 805, 809, 810, 811, 813, 817, 820, 821, 826, 830
  Summary, 716, 752, 792, 834
  Understand, 696, 704, 713, 732, 740, 749, 754, 766, 774, 782, 789, 806, 814, 822, 831, 836
Kuiper belt, 783

L

Lab, 714–715, 750–751, 790–791, 832–833. *See also* **Launch Lab; MiniLab; Skill Practice**
Lander(s), 701
Launch Lab, 689, 699, 707, 725, 735, 743, 761, 769, 777, 785, 801, 809, 817, 825
Lens, 689 *lab*
Lesson Review, 696, 704, 713, 732, 740, 749, 766, 774, 782, 806, 814, 822, 831
Light
  speed of, 691
  types of, 803 *lab*
  white, 691 *lab*
Long-period comet(s), 788
Luminosity
  explanation of, **805**
  of stars, 813, *813*
Lunar, 701
Lunar cycle, 738, *739*
Lunar eclipse(s)
  explanation of, **746**
  occurrence of, 747
  partial, 747, *747*
  total, 746, *746*
Lunar probe(s), 701, 707
Lunar Reconnaissance Orbiter (LRO), 707, 741

M

***Magellan* space probe,** 771
Makemake
  as dwarf planet, 785, 786
  explanation of, 763
Mantle
  of Earth, 772
March Equinox, *730,* 731, *731*
Maria, 736, *737*
***Mariner* 688, 696,** *700,* 701, 708
Mars, *706*
  atmosphere on, 773
  data on, *773*
  explanation of, 708, 763, *769,* 773
  exploration of, 710
  surface of, 773
Mars Science *Laboratory,* 711
Math Skills, 691, 696, 729, 732, 755, 779, 782, 795, 804, 806, 837
Mercury
  data for, *770*
  explanation of, 708, 763, *769,* 770
  structure and surface of, 770
Messenger, 708, *708*
Meteor(s), 767, **788**
Meteorite(s), 767, **788**
Meteoroid(s), 788
Methane gas
  on Uranus, 780, *780*
Milky Way galaxy, 689, 827, *828–829*
MiniLab, 691, 703, 710, 726, 738, 744, 765, 772, 781, 788, 803, 810, 820, 827. *See also* **Lab**
Moon(s)
  Apollo Space Program on, 741
  appearance of, 735 *lab*
  data on, *735*
  effect on tides by, 748, *748*
  explanation of, 700, 735
  exploration of, 710, *710,* 714 *lab*
  formation of, 736, *736, 737,* 785 *lab*
  of Jupiter, 778
  lunar eclipses and, 746, 746–747, *747*
  phases of, 738, *739,* 750–751 *lab*
  probes to, 701, 707
  rotation of, 737, *737,* 738, 738 *lab,* 744, *745*
  of Saturn, 780, *780,* 781 *lab*
  solar eclipses and, 744, *744,* 745, *745*
  surface of, 736
  of Uranus, 781

N

Naked-eye astronomy, 802
NASA
  space program of, 741
National Aeronautics and Space Administration (NASA), 700, 703, 710, 711, 815
Neap tide(s), 748
Nebula, 817
Neptune
  data on, *781*
  explanation of, 709, 763, 781
Neutron star, 820
New Horizons, 709, *709*
Nuclear fusion, 726, **809**

O

Olympus Mons, 773, *773*
Open cluster(s), 812, *812*
Opportunity, 708, *708*
Optical space telescope(s), 692, *692,* 694, *694*
Option, 710

Orbit

Orbit
of Earth around Sun, 726, *726*, 726 *lab*, 727, *727*
elliptical, 765, 765 *lab*
explanation of, **726**, 762
Orbiter(s), *701*, 709
Outer planet(s). *See also* **Planets**; *specific planets*
explanation of, *762*, 763, 777, *777*
Jupiter, *778*, 778–779
Neptune, 781, *781*
Saturn, *779*, 779–780, *780*
Uranus, *780*, 780–781

P

Parallax, 804
Penumbra, *743*, **743**
Period of revolution, 764
Period of rotation, 764
Phase, 738
Photosphere, 810
Pioneer *696*, *700*
Planet(s). *See also* **Inner planet(s)**; **Outer planet(s)**; *specific planets*
distance from Sun of, *764*
dwarf, 709, *785*, 785–786, *786*
explanation of, 3
formation of, 767
graphing characteristics of, 775
inner and outer, 762, 763
motion of, 764–765, *765*
orbits and speeds of, 765, *765*
space missions to inner, 708, *708*
Planetary nebulae, 821, *821*
Pluto
as dwarf planet, 783, 785, 786, *786*
explanation of, 763
Polaris, 801, *801*
Probe, 779
Project Apollo, 702
Prominence, *811*
Ptolemy, 802

R

Radiant energy, 690
Radiative zone, 810
Radio telescope(s), *693*, 693
Radio wave(s), 693
Ray(s)
from Moon craters, 736, *737*
Reading Check, , 689, 690, 691, 693, 695, 702, 708, 709, 727, 736, 761, 762, 764, 772, 786, 804, 812, 819, 825, 827
Red giant
Sun as, 819, *819*
Reflecting telescope(s), *692*, **692**
Refracting telescope(s), *692*, **692**
Review Vocabulary, 709, 762, 820. *See also* **Vocabulary**
Revolution
explanation of, **726**
period of, 764
Rocket(s), *698*
explanation of, **699**, 699 *lab*, 705

Rotation
of Earth around Sun, 726, *726*, 726 *lab*, 727, *727*
explanation of, **727**
of Moon around Earth, 737, *737*, 738
period of, **764**
Rotation axis
explanation of, *727*, **727**, 729
seasons and Earth's, 729, *729*, 731, *731*, 733

S

Satellite(s). *See also* Space exploration
explanation of, **700**
that orbit Earth, 712, *712*
Saturn
data on, *779*
explanation of, 709, 763, 779
moons of, 780, *780*, 781 *lab*
structure of, 779
Science Methods, 833
Science & Society, 715, 741, 751, 791
Science Use v. Common Use, 701, 738, 762, 809. *See also* Vocabulary
Season(s)
Earth's rotational axis and, 729, *729*, 731, *731*, 733
explanation of, 729
September equinox, *730*, 731, *731*
Shadow(s)
changes in, 743 *lab*
explanation of, 743
Short-period comets, 788
Skill Practice, 697, 733, 775, 807, 823. *See also* **Lab**
Sky
appearance of night, 801, *801*, 802
methods to view, 801–803
observation of, 689, 802
Solar eclipse(s)
explanation of, **744**, *744*, 745
occurrence of, 745, *745*
partial, 745
total, 744
Solar nebula, 767
Solar probe(s), 707
Solar system. *See also* **Planets**; *specific planets*
early exploration of, 700, *700*
explanation of, 3, *3*
measuring distances in, 804, *804*
modeling of, 790–791 *lab*
motions of planets in, 764–765, *765*
objects in, *762*, 762–764, *763*, 764
observation of, 777 *lab*
water in, 711, *711*
Solar Terrestrial Relations Observation (STEREO) telescopes, 815
Solar wind, *811*
Solstice
December, *730*, 731, *731*
explanation of, **731**
June, *730*, 731, *731*
Sound wave(s), 690

Supernovae

Space elevator(s), 705, *705*
Space exploration
events in early, 700, *700*
humans in, 702, *702*, 710, *710*
to inner planets, 708, *708*
insight into Earth from, 712
to outer planers, 709, *709*
technology for, 703
Space probe(s)
explanation of, **701**, *701*
to sun and moon, 707
Space shuttle(s), 699, 702, *702*
Spectroscope(s), 803
Spiral galaxy(ies), *826*, 827
Spirit, 708, *708*
Spitzer Space Telescope, 695
Spring
in Northern Hemisphere, 729, *729*
Spring tide(s), 748
Sputnik **687,** 700
Standardized Test Practice, 720–721, 756–757, 796–797, 838–839
Star(s)
appearance of, 801, *801*
brightness of, 805, *805*
classification of, *812*, 812–813
composition and structure of, 809–810
end of, *819*, 819–821
explanation of, 761, **762**, 809
groups of, *812*, 812
life cycle of, 817–818, *818*
light from, 691
main-sequence, 813, 818
radiant energy and, 690
Star clusters, *812*, 812
Stellar, 809
Study Guide, 716, 752–753, 792–793, 834–835
Summer
in Northern Hemisphere, 729, *729*
Sun, 707
apparent motion of, 727
changing features of, 810, *811*, 819, *819*
distance between Earth and, 726
Earth's orbit around, 726, *726*, 726 *lab*, 727, *727*
effect on tides by, 748, *748*
energy from, 726, *726*, 728, *728*
energy produced by, 801 *lab*
explanation of, 689, 725, *725*
diameter of, 762
distance of planets from, *764*
gravitational force of, 764, *765*
layers of, *810*
objects that orbit, *762*, 762–763, *763*, 765
probes to, 707
seasons on Earth and, 729, *729*, 731, *731*, 733
spots on, 809 *lab*
structure of, 810 *lab*
Sunspot(s), *811*
Supernovae, *819*, 821

I-4 • Index

T

Technology
  space, 703, 703
Telescope(s)
  construction of simple, 697, *697*
  explanation of, 689, 803
  optical, 692, *692*, 694
  optical space, 694, *694*
  radio, 693, *693*
  Solar Terrestrial Relations Observatory, 815
  space, 694, *694*
Temperature
  of stars, 812, *812*, 813, *813*
Terrestrial, 769
Terrestrial planet(s). *See also* **Inner planet(s)**
  explanation of, **769**
Tide(s)
  explanation of, 747
  Moon's effect on, 748, *748*
  Sun's effect on, 748, *748*
Tyson, Neil deGrasse, 783

U

Umbra, *743*, **743**
Universe
  movement in, 825 *lab*
  origin and expansion of, 830
  traveling through, 832–833 *lab*
Uranus
  axis and moons of, 781
  data on, *780*
  explanation of, 709, 763, 780
Ursa Major, 807

V

Valles Marineris, 773, *773*
Venus
  atmosphere of, 771
  data on, *771*
  explanation of, 708, 763, *769*, 771
  greenhouse effect on, 771
  structure and surface of, 771, *771*
Visual Check, 690, 694, 700, 701, 709, 712, 728, 731, 744, 745, 746, 763, 769, 777, 786, 802, 803, 805, 810, 813, 818
Vocabulary, 688, 698, 706, 723, 724, 734, 752, 759, 760, 768, 776, 784, 792, 800, 806, 808, 816, 824. *See also* **Academic Vocabulary; Review Vocabulary; Science Use v. Common Use; Word Origin**
  Use, 696, 704, 713, 717, 732, 740, 749, 753, 766, 774, 782, 789, 793, 806, 814, 822, 831, 835
Voyager 780, **760**

W

Waning phase(s), 738
Water
  in solar system, 711, 711
Wavelength(s), 803, *803*
Waxing phase(s), 738
What do you think?, 687, 696, 704, 713, 723, 732, 740, 749, 759, 766, 774, 782, 789, 799, 806, 814, 822, 831
White dwarf, *819,* **819,** 821, *821*
White light, 691 *lab*
Winter
  in Northern Hemisphere, 729, *729*
Word Origin, 689, 700, 711, 731, 736, 743, 763, 769, 780, 788, 812, 817, 827. *See also* **Vocabulary**
Writing In Science, 719, 755, 795, 837

Credits

Photo Credits

COV K. R. Svensson/Photo Researchers; **ii** K. R. Svensson/Photo Researchers; **vii** Ransom Studios; **viii** Daniel H. Bailey/Alamy; **ix** ©Fancy Photography/Veer; **684** (t)courtesy of PowerFilm, Inc., (c)courtesy of Seldon Technologies, (b)Jacques Descloitres, MODIS Rapid Response Team, NASA/GSFC; **685** (t) USGS, (b)NOAA; **686–687** Stocktrek/age fotostock; **688** NASA and The Hubble Heritage Team (AURA/STScI); **689** Hutchings Photography/Digital Light Source; **691** Hutchings Photography/Digital Light Source; **692** (inset) Richard Wainscoat/Alamy, (bkgd)Roger Ressmeyer/Corbis; **693** (t)Images Etc Ltd/Getty Images, (b)Starfire Optical Range/USAF/Roger Ressmeyer/Corbis, (inset)Time & Life Pictures/Getty Images; **694** NASA; **695** (t)NASA/JPL-Caltech/STScI/CXC/SAO, (b)NASA/GSFC; **696** (t)Time & Life Pictures/Getty Images, (bl br)NASA; **697** (1 6 7)Digital Light Source, (others) McGraw-Hill Education; **698** Stockbyte/Alamy; **699** (t)Hutchings Photography/Digital Light Source, (b)©Stocktrek/age fotostock; **700** (l) NASA Marshall Space Flight Center (NASA-MSFC), (cl)NASA, (cr)NASA/JPL, (r)Stocktrek/Corbis; **701** (l)AP Images, (c)NASA/JPL, (r)Atlas Photo Bank/Photo Researchers, Inc; **702** (l r)NASA, (c)Stocktrek/age fotostock; **703** (t) Hutchings Photography/Digital Light Source, (b)Alex Bartel/Photo Researchers, Inc.; **704** (t)©Stocktrek/age fotostock, (c)NASA/JPL, (bl)Alex Bartel/Photo Researchers, Inc., (br)Hutchings Photography/Digital Light Source; **706** NASA/JPL/University of Arizona; **707** (t)Hutchings Photography/Digital Light Source, (b)SOHO (NASA & ESA); **708** (l)NASA/Johns Hopkins University Applied Physics Laboratory/Carnegie Institution of Washington., (r)©NASA/epa/Corbis; **709** (l)Craig Attebery/NASA, (r)NASA/Johns Hopkins University Applied Physics Laboratory/Southwest Research Institute (NASA/JHUAPL/SwRI); **710** (t)Michael Hixenbaugh/National Science Foundation, (b)Hutchings Photography/Digital Light Source; **711** (t) Arco Images GmbH/Alamy, (bl)NASA/JPL/DLR, (br)NASA/JPL/University of Arizona/University of Colorado; **712** (t)NASA/Ames Wendy Stenzel, (b) NASA/Goddard Space Flight Center Scientific Visualization Studio; **713** (t r) NASA/Johns Hopkins University Applied Physics Laboratory/Southwest Research Institute (NASA/JHUAPL/SwRI), (c)NASA/JPL/University of Arizona/University of Colorado, (b)NASA/Goddard Space Flight Center Scientific Visualization Studio; **714** (r)Hutchings Photography/Digital Light Source, (others)McGraw-Hill Education; **716** (t)NASA, (c)©Stocktrek/age fotostock, (b)Michael Hixenbaugh/National Science Foundation; **718** NASA/JPL/Cornell; **719** Stocktrek/age fotostock; **722–723** O. Alamany & E. Vicens/Corbis; **724** NASA Human Spaceflight Collection; **725** (t)Hutchings Photography/Digital Light Source, (b)SOHO (ESA & NASA); **726** Hutchings Photography/Digital Light Source; **732** SOHO (ESA & NASA); **733** Hutchings Photography/Digital Light Source; **734** NASA; **735** Hutchings Photography/Digital Light Source; **737** (c)NASA/JPL/USGS, (bl)ClassicStock/Alamy, (others)Lunar and Planetary Institute; **738** Hutchings Photography/Digital Light Source; **739** Eckhard Slawik/Photo Researchers, Inc.; **740** (c)Lunar and Planetary Institute, (b)Eckhard Slawik/Photo Researchers, Inc.; **741** NASA; **742** Jacques Descloitres, MODIS Rapid Response Team at NASA GSFC; **743** Hutchings Photography/Digital Light Source; **744** Hutchings Photography/Digital Light Source; **747** Robert Estall photo agency/Alamy; **750** (2) McGraw-Hill Education, (others)Hutchings Photography/Digital Light Source; **751** (t to b)NASA, m-gucci/Getty Images, Brian E. Kushner/Getty Images; **752** Eckhard Slawik/Photo Researchers, Inc.; **755** O. Alamany & E. Vicens/Corbis; **758–759** NASA/JPL/Space Science Institute; **760** UVimages/amanaimages/Corbis; **761** (t)Hutchings Photography/Digital Light Source, (b)Diego Barucco/Alamy; **763** NASA/JPL; **765** Hutchings Photography/Digital Light Source; **766** UVimages/amanaimages/Corbis; **767** (t)Josef Muellek/Getty Images, (c)American Museum of Natural History; **768** ESA/DLR/FU Berlin (G. Neukum); **769** (t)Hutchings Photography/Digital Light Source, (bl)NASA/Johns Hopkins University Applied Physics Laboratory/Carnegie Institution of Washington, (bcl)NASA, (bcr)NASA Goddard Space Flight Center, (br)NASA/JPL/Malin Space Science Systems; **770** NASA/Johns Hopkins University Applied Physics Laboratory/Carnegie Institution of Washington; **771** NASA/JPL; **772** (tl)NASA, (r)Image Ideas/PictureQuest, (bl)Comstock/JupiterImages; **773** (tl)NASA/JPL, (b)NASA/JPL/University of Arizona; **774** (t)NASA/Johns Hopkins University Applied Physics Laboratory/Carnegie Institution of Washington, (c)Comstock/JupiterImages, (bl)NASA; **776** NASA/JPL; **777** (l)NASA/JPL/USGS, (cl)NASA and The Hubble Heritage Team (STScI/AURA)Acknowledgment: R.G. French (Wellesley College), J. Cuzzi (NASA/Ames), L. Dones (SwRI), and J. Lissauer (NASA/Ames), (cr r) NASA/JPL; **778** NASA/JPL; **780** (bl)NASA/ESA and Erich Karkoschka, University of Arizona, (others)NASA/JPL/Space Science Institute; **781** NASA/JPL; **782** (t b)NASA/JPL, (c)NASA/JPL/USGS; **783** Frederick M. Brown/Getty Images; **784** Gordon Garradd/SPL/Photo Researchers, Inc.; **785** Hutchings Photography/Digital Light Source; **786** (l)Dr. R. Albrecht, ESA/ESO Space Telescope European Coordinating Facility; NASA, (tr)NASA, ESA, and J. Parker (Southwest Research Institute), (br)NASA, ESA, and M. Brown (California Institute of Technology); **787** (1 2)NASA/JPL/JHUAPL, (3)NASA/JPL/USGS, (4)Ben Zellner (Georgia Southern University), Peter Thomas (Cornell University), NASA/ESA, (5)Roger Ressmeyer/Photographer's Choice/Getty Images, (6)NASA/JPL-Caltech; **788** (t)Jonathan Blair/Corbis, (b)Hutchings Photography/Digital Light Source; **789** (t)NASA/JPL/USGS, (c)Gordon Garradd/SPL/Photo Researchers, Inc., (bl)Jonathan Blair/Corbis, (br)Roger Ressmeyer/Photographer's Choice/Getty Images; **790** Hutchings Photography/Digital Light Source; **791** Hutchings Photography/Digital Light Source; **794** NASA Goddard Space Flight Center; **795** NASA/JPL/Space Science Institute; **798–799** NASA, ESA, and S. Beckwith (STScI)and the HUDF Team; **800** Stephen & Donna O'Meara/Photo Researchers, Inc.; **801** (t) Hutchings Photography/Digital Light Source, (b)Joseph Baylor Roberts/Getty Images; **803** (tl)Robert Gendler/Stocktrek Images/Getty Images, (tr) NASA/CXC/MIT/H. Marshall et al., (cl)NASA/JPL-Caltech/E. Churchwell (University of Wisconsin), (cr)NASA/JPL-Caltech/Univ. of Virginia, (b) Hutchings Photography/Digital Light Source; **807** (t to b)McGraw-Hill Education, Brand X Pictures/PunchStock, Aaron Haupt, Hutchings Photography/Digital Light Source; **808** Science Source/Photo Researchers, Inc; **809** Digital Vision/PunchStock; **810** Hutchings Photography/Digital Light Source; **811** (t to b)Jerry Lodriguss/Photo Researchers, Inc., Naval Research Laboratory, SOHO Consortium, ESA, NASA, Arctic-Images/Getty Images; **812** (t)Heidi Schweiker/WIYN and NOAO/AURA/NSF, (b)NOAO/AURA/NSF; **814** (t)Jerry Lodriguss/Photo Researchers, Inc., (br)NOAO/AURA/NSF; **815** STEREO Stereoscopic Observations Constraining the Initiation of Polar Coronal Jets S. Patsourakos, E.Pariat, A. Vourlidas, S. K. Antiochos, J. P. Wuesler/NASA; **816** NASA; **817** Hutchings Photography/Digital Light Source; **820** Hutchings Photography/Digital Light Source; **821** (t)X-ray:

PERIODIC TABLE OF THE ELEMENTS

Key:
- Element — Hydrogen
- Atomic number — 1
- Symbol — H
- Atomic mass — 1.01
- State of matter

Legend:
- Gas
- Liquid
- Solid
- Synthetic

A column in the periodic table is called a **group**.

A row in the periodic table is called a **period**.

Period	1	2	3	4	5	6	7	8	9
1	Hydrogen 1 H 1.01								
2	Lithium 3 Li 6.94	Beryllium 4 Be 9.01							
3	Sodium 11 Na 22.99	Magnesium 12 Mg 24.31							
4	Potassium 19 K 39.10	Calcium 20 Ca 40.08	Scandium 21 Sc 44.96	Titanium 22 Ti 47.87	Vanadium 23 V 50.94	Chromium 24 Cr 52.00	Manganese 25 Mn 54.94	Iron 26 Fe 55.85	Cobalt 27 Co 58.93
5	Rubidium 37 Rb 85.47	Strontium 38 Sr 87.62	Yttrium 39 Y 88.91	Zirconium 40 Zr 91.22	Niobium 41 Nb 92.91	Molybdenum 42 Mo 95.96	Technetium 43 Tc (98)	Ruthenium 44 Ru 101.07	Rhodium 45 Rh 102.91
6	Cesium 55 Cs 132.91	Barium 56 Ba 137.33	Lanthanum 57 La 138.91	Hafnium 72 Hf 178.49	Tantalum 73 Ta 180.95	Tungsten 74 W 183.84	Rhenium 75 Re 186.21	Osmium 76 Os 190.23	Iridium 77 Ir 192.22
7	Francium 87 Fr (223)	Radium 88 Ra (226)	Actinium 89 Ac (227)	Rutherfordium 104 Rf (267)	Dubnium 105 Db (268)	Seaborgium 106 Sg (271)	Bohrium 107 Bh (272)	Hassium 108 Hs (270)	Meitnerium 109 Mt (276)

The number in parentheses is the mass number of the longest lived isotope for that element.

Lanthanide series

Cerium 58 Ce 140.12	Praseodymium 59 Pr 140.91	Neodymium 60 Nd 144.24	Promethium 61 Pm (145)	Samarium 62 Sm 150.36	Europium 63 Eu 151.96

Actinide series

Thorium 90 Th 232.04	Protactinium 91 Pa 231.04	Uranium 92 U 238.03	Neptunium 93 Np (237)	Plutonium 94 Pu (244)	Americium 95 Am (243)

Credits

NASA/CXC/SAO; Optical: NASA/STScI, (b)NASA, The Hubble Heritage Team (STScI/AURA), Y.-H. Chu (UIUC), S. Kulkarni (Caltech)and R. Rothschild (UCSD); **822** (l)NASA, The Hubble Heritage Team (STScI/AURA), Y.-H. Chu (UIUC), S. Kulkarni (Caltech)and R. Rothschild (UCSD), (r)NASA; **823** Aaron Haupt; **824** NASA/Hubble Heritage Team; **825** NASA/Alamy; **826** (t b)NASA, ESA, M. Livio and the Hubble Heritage Team (STScI/AURA), (c)Robert Gendler/NASA; **827** STS-82 Crew/STScI/NASA; **831** NASA/Alamy; **832** (1 4) Hutchings Photography/Digital Light Source, (others)McGraw-Hill Education; **833** (t)Hutchings Photography/Digital Light Source, (b)NASA/JPL-Caltech, (inset)Brand X Pictures/PunchStock, (inset)NASA/JPL-Caltech/S. Willner (Harvard-Smithsonian Center for Astrophysics); **836** Robert Gendler/NASA; **839** NASA, ESA, and The Hubble Heritage Team (STScI/AURA); **SR-0–SR-1** Gallo Images - Neil Overy/Getty Images; **SR-2** Hutchings Photography/Digital Light Source; **SR-6** Michell D. Bridwell/PhotoEdit; **SR-7** (t)McGraw-Hill Education, (b)Dominic Oldershaw; **SR-8** StudiOhio; **SR-9** Timothy Fuller; **SR-10** Aaron Haupt; **SR-12** KS Studios; **SR-13** Matt Meadows.